Research Notes in Mathematics

Submission of proposals for consideration
Suggestions for publication, in the form of outlines and representative samples, are invited by the Editorial Board for assessment. Intending authors should approach one of the main editors or another member of the Editorial Board, citing the relevant AMS subject classifications. Alternatively, outlines may be sent directly to one of the publisher's offices. Refereeing is by members of the board and other mathematical authorities in the topic concerned, throughout the world.

Preparation of accepted manuscripts
On acceptance of a proposal, the publisher will supply full instructions for the preparation of manuscripts in a form suitable for direct photo-lithographic reproduction. Specially printed grid sheets are provided and a contribution is offered by the publisher towards the cost of typing. Word processor output, subject to the publisher's approval, is also acceptable.

Illustrations should be prepared by the authors, ready for direct reproduction without further improvement. The use of hand-drawn symbols should be avoided wherever possible, in order to maintain maximum clarity of the text.

The publisher will be pleased to give any guidance necessary during the preparation of a typescript, and will be happy to answer any queries.

Important note
In order to avoid later retyping, intending authors are strongly urged not to begin final preparation of a typescript before receiving the publisher's guidelines and special paper. In this way it is hoped to preserve the uniform appearance of the series.

Advanced Publishing Program
Pitman Publishing Inc
1020 Plain Street
Marshfield, MA 02050, USA
(tel (617) 837 1331)

Advanced Publishing Program
Pitman Publishing Limited
128 Long Acre
London WC2E 9AN, UK
(tel 01-379 7383)

Titles in this series

1 Improperly posed boundary value problems
A Carasso and A P Stone

2 Lie algebras generated by finite dimensional ideals
I N Stewart

3 Bifurcation problems in nonlinear elasticity
R W Dickey

4 Partial differential equations in the complex domain
D L Colton

5 Quasilinear hyperbolic systems and waves
A Jeffrey

6 Solution of boundary value problems by the method of integral operators
D L Colton

7 Taylor expansions and catastrophes
T Poston and I N Stewart

8 Function theoretic methods in differential equations
R P Gilbert and R J Weinacht

9 Differential topology with a view to applications
D R J Chillingworth

10 Characteristic classes of foliations
H V Pittie

11 Stochastic integration and generalized martingales
A U Kussmaul

12 Zeta-functions: An introduction to algebraic geometry
A D Thomas

13 Explicit *a priori* inequalities with applications to boundary value problems
V G Sigillito

14 Nonlinear diffusion
W E Fitzgibbon III and H F Walker

15 Unsolved problems concerning lattice points
J Hammer

16 Edge-colourings of graphs
S Fiorini and R J Wilson

17 Nonlinear analysis and mechanics: Heriot-Watt Symposium Volume I
R J Knops

18 Actions of fine abelian groups
C Kosniowski

19 Closed graph theorems and webbed spaces
M De Wilde

20 Singular perturbation techniques applied to integro-differential equations
H Grabmüller

21 Retarded functional differential equations: A global point of view
S E A Mohammed

22 Multiparameter spectral theory in Hilbert space
B D Sleeman

24 Mathematical modelling techniques
R Aris

25 Singular points of smooth mappings
C G Gibson

26 Nonlinear evolution equations solvable by the spectral transform
F Calogero

27 Nonlinear analysis and mechanics: Heriot-Watt Symposium Volume II
R J Knops

28 Constructive functional analysis

29 Elongational flows: Aspects of the behaviour of model elasticoviscous fluids
C J S Petrie

30 Nonlinear analysis and mechanics: Heriot-Watt Symposium Volume III
R J Knops

31 Fractional calculus and integral transforms of generalized functions
A C McBride

32 Complex manifold techniques in theoretical physics
D E Lerner and P D Sommers

33 Hilbert's third problem: scissors congruence
C-H Sah

34 Graph theory and combinatorics
R J Wilson

35 The Tricomi equation with applications to the theory of plane transonic flow
A R Manwell

36 Abstract differential equations
S D Zaidman

37 Advances in twistor theory
L P Hughston and R S Ward

38 Operator theory and functional analysis
I Erdelyi

39 Nonlinear analysis and mechanics: Heriot-Watt Symposium Volume IV
R J Knops

40 Singular systems of differential equations
S L Campbell

41 N-dimensional crystallography
R L E Schwarzenberger

42 Nonlinear partial differential equations in physical problems
D Graffi

43 Shifts and periodicity for right invertible operators
D Przeworska-Rolewicz

44 Rings with chain conditions
A W Chatters and C R Hajarnavis

45 Moduli, deformations and classifications of compact complex manifolds
D Sundararaman

46 Nonlinear problems of analysis in geometry and mechanics
M Atteia, D Bancel and I Gumowski

47 Algorithmic methods in optimal control
W A Gruver and E Sachs

48 Abstract Cauchy problems and functional differential equations
F Kappel and W Schappacher

49 Sequence spaces
W H Ruckle

50 Recent contributions to nonlinear partial differential equations
H Berestycki and H Brezis

51 Subnormal operators
J B Conway

52 Wave propagation in viscoelastic media
F Mainardi

53 Nonlinear partial differential equations and their applications: Collège de France Seminar. Volume I
H Brezis and J L Lions

54 Geometry of Coxeter groups
H Hiller

55 Cusps of Gauss mappings
T Banchoff, T Gaffney and C McCrory

106 Nonlinear partial differential equations and free boundaries. Volume I
J I Díaz

107 Tight and taut immersions of manifolds
T E Cecil and P J Ryan

108 A layering method for viscous, incompressible L_p flows occupying R^n
A Douglis and E B Fabes

109 Nonlinear partial differential equations and their applications: Collège de France Seminar. Volume VI
H Brezis and J L Lions

110 Finite generalized quadrangles
S E Payne and J A Thas

111 Advances in nonlinear waves. Volume II
L Debnath

112 Topics in several complex variables
E Ramírez de Arellano and D Sundararaman

113 Differential equations, flow invariance and applications
N H Pavel

114 Geometrical combinatorics
F C Holroyd and R J Wilson

115 Generators of strongly continuous semigroups
J A van Casteren

116 Growth of algebras and Gelfand–Kirillov dimension
G R Krause and T H Lenagan

117 Theory of bases and cones
P K Kamthan and M Gupta

118 Linear groups and permutations
A R Camina and E A Whelan

119 General Wiener–Hopf factorization methods
F-O Speck

120 Free boundary problems: applications and theory, Volume III
A Bossavit, A Damlamian and M Fremond

121 Free boundary problems: applications and theory, Volume IV
A Bossavit, A Damlamian and M Fremond

122 Nonlinear partial differential equations and their applications: Collège de France Seminar. Volume VII
H Brezis and J. L. Lions

123 Geometric methods in operator algebras
H Araki and E G Effros

124 Infinite dimensional analysis–stochastic processes
S Albeverio

125 Ennio de Giorgi Colloquium
P Krée

126 Almost-periodic functions in abstract spaces
S Zaidman

127 Nonlinear variational problems
A Marino, L Modica, S Spagnolo and M Degiovanni

128 Second-order systems of partial differential equations in the plane
L K Hua, W Lin and C-Q Wu

129 Asymptotics of high-order ordinary differential equations
R B Paris and A D Wood

130 Stochastic differential equations
R Wu

Paul Krée (Editor)

Université Pierre et Marie Curie (Paris VI)

Ennio de Giorgi Colloquium

π

Pitman Advanced Publishing Program

BOSTON · LONDON · MELBOURNE

PITMAN PUBLISHING INC
1020 Plain Street, Marshfield, Massachusetts 02050

PITMAN PUBLISHING LIMITED
128 Long Acre, London WC2E 9AN

Associated Companies
Pitman Publishing Pty Ltd, Melbourne
Pitman Publishing New Zealand Ltd, Wellington
Copp Clark Pitman, Toronto

First published 1985

AMS Subject Classifications: 35-06, 49-06

ISSN 0743-0337

Library of Congress Cataloging in Publication Data

Ennio de Giorgi Colloquium (L983 : H. Poincaré Institute)
Ennio de Giorgi Colloquium.

Papers presented at a colloquium held at the H. Poincaré Institute in Nov. 1983.
Bibliography: p.
1. Differential equations, Partial—Congresses.
2. Surfaces, Minimal—Congresses. 3. De Giorgi, Ennio—Congresses. I. Krée, Paul.
QA377.E56 1983 515.3′53 84-26536
ISBN 0-273-08680-4

British Library Cataloguing in Publication Data

Ennio de Giorgi Colloquium *(H. Poincaré Institute : 1983)*
Ennio de Giorgi Colloquium.—(Research notes in mathematics, ISSN 0743-0337; 125)
1. Calculus
I. Title II. Krée, Paul III. Series
515 QA303

ISBN 0-273-08680-4

Reproduced and printed by photolithography
in Great Britain by Biddles Ltd, Guildford

Contents

Preface

On 4th July 1983, Professor Ennio de GIORGI was awarded the title "Doctor Honoris Causa" by the Council of the Université de PARIS VI. It was announced in September 1983 that the ceremony would take place at the Sorbonne on the 7th November, in the presence of the Rector and of a large invited audience of ambassadors and scientists.

The very profound and influential nature of Ennio de Giorgi's work meant that this award had a considerable impact on the international scientific community, particularly in France and Italy. The happy result was that, although only a short time was available for its organization, a French-Italian colloquium centering on de Giorgi's work was held at the H. Poincaré Institute on the 4th, 5th and 7th November, 1983.

The scientific committee for this colloquium was the following :

J. LERAY, Président - H. BREZIS - P. DOLBEAULT - M. HERVE
P. KREE (Secretary) - J.L. LIONS - P. MALLIAVIN and J. VAILLANT.

This scientific committee thanks the sixteen lecturers for their contributions and is also grateful to the 200 mathematicians who participated in the meeting. Particular thanks are due to Jean ASTIER, President of the Université de Paris VI, who helped us very efficiently with its organization, and to the following organizations who gave financial support :

Présidence de l'Université de PARIS VI, Société Mathématique de France, Ambassade de France à Rome, Ministère des Relations extérieures, Département de Mathématique (Paris VI). The administrative help of Daniel GABAY, Scientific Attaché to the French Embassy in Rome, is also particularly appreciated.

This Research Note includes sixteen papers reporting actual mathematical research in France and Italy related to the work of Ennio de GIORGI. In order to help in tracing references, de Giorgi's scientific work is briefly summarized and a bibliography is also given.

The scientific committee thanks Anne-Marie DURRANDE (Secretary of the Department of Mathematics, Université de PARIS VI) for her help, Mme NOTAIRE for her excellent typing, and Pitman Publishing who proposed and realized the international edition of the Ennio de Giorgi colloquium.

March 1985 Paul KREE

J LERAY

Opening address

Ladies and Gentlemen :

We meet to honour the worldwide impact of Italian mathematics, and, particularly, one of its leaders, Professor Ennio DE GIORGI, Corresponding Member of the Accademia Nationale dei Lincei, Member of the Accademia Nazionale Delta dei XL, of the Academia Pontificia Scientiarum, of the Accademia delle Scienze di Torino, of the Istituto Lombardo Accademia Scienze e Lettere and of the Scientific Academy of Latin America. He and several of his many collaborators will present papers. Of course, the main centre of their activity is the Scuola Normale Superiore di Pisa, and often they are the guests of foreign institutes. It is, however, quite exceptional that so many of them meet far away: the Université de Paris is proud to be their gathering place today.

It would be unrealistic to represent de Giorgi's school in few words, impossible to describe the rôles played by those of its members who are present here or even to try to summarize de Giorgi's work. However, the Proceedings of this meeting contain a list of his papers - actually a very incomplete list: it cannot quote his future publications !

If de Giorgi's work cannot be briefly analyzed, it can easily be characterized: a continuous renewal, an always surprising profundity, a great beauty. This dazzling scientist is also high-minded.

And the best to do is to let him and his colleagues say as soon as possible what they have come to say.

Jean LERAY

E DE GIORGI

Some semi-continuity and relaxation problems

In this paper I present some recent results and some still open questions concerning the study of semicontinuity and relaxation problems in the Calculus of Variations. From the problem presented here it seems possible to conclude that the semicontinuity and relaxation theory, although widely studied, remains a field of research still open and rich in interesting problems.

To fix the terms of this discussion I recall briefly the main definitions of the theory of semicontinuity and relaxation.

Let X be a topological space, $\overline{\mathbb{R}} = \mathbb{R} \cup \{-\infty, +\infty\}$, $E \subseteq X$ and let $F : E \to \overline{\mathbb{R}}$ be a function. We shall indicate by $\Gamma(X^-)F$ the function from the closure $\overline{E}$ of E into $\overline{\mathbb{R}}$ defined by

$$\Gamma(X^-)F(X) = \liminf_{y \to x} F(y) = \sup_{A \in I(x)} \inf_{y \in E \cap A} F(y),$$

where $I(x)$ denotes the family of neighbourhoods of x in X.

The following proposition is immediate :

<u>Proposition 1</u>

> We have $\inf_E F = \inf_{\overline{E}} \Gamma(X^-)F$. Moreover, if F is coercive on X, that is
>
> $$\forall \lambda \in \mathbb{R} \ \exists K_\lambda \text{ compact subset of } X : \{x \in E : F(x) \leq \lambda\} \subseteq K_\lambda ,$$
>
> then we have $\inf_E F = \min_{\overline{E}} \Gamma(X^-)F$.

In many problems in the Calculus of Variations, given a functional of the form

$$F(u) = \int_\Omega f(x,u,Du,\ldots,D^m u)\, dx ,$$

it is of interest to study the following problems :

PROBLEM 1 - Semicontinuity problem

Find conditions under which the functional F is lower semicontinuous with respect to the $L^1_{loc}(\Omega)$ -topology.

PROBLEM 2 - Relaxation problem

If the functional F is not $L^1_{loc}(\Omega)$ -lower semicontinuous, find conditions under which we have

$$\Gamma(L^1_{loc}(\Omega)^-) F(u) = \int_\Omega g(x,u,Du,\dots D^m u)\, dx$$

for some suitable integrand g (and if possible characterize this new integrand).

Problem 1, in the case $F(u) = \int_\Omega f(x,u,Du)\, dx$, has been studied by many authors (see for example [2], [8], [12], [13], [15]) when $f(x,s,z)$ is a function satisfying the usal Caratheodory hypotheses (i.e. measurable in x and continuous in (s,z)); Problem 2 was discussed later under similar conditions (see for example [1], [11], [12]).

The hypothesis of continuity with respect to s has been weakened by some authors by requiring only the lower semicontinuity with respect to s (see for example [6], [14]). But, as far as I know, the case when $f(x,s,z)$ is only measurable with respect to s has not received much attention. A recent result in this direction, concerning functionals of the form $\int_\Omega f(u,Du)\, dx$, is the following.

Theorem 1 (see [9]).

> Let $f : \mathbb{R} \times \mathbb{R}^n \to [0,+\infty[$ be a function with the following properties
>
> a) for every $z \in \mathbb{R}^n$ the function $f(\cdot,z)$ is measurable ;
>
> b) for every $s \in \mathbb{R}^n$ the function $f(s,\cdot)$ is convex ;
>
> c) the function $f(\cdot,0)$ is lower semicontinuous ;
>
> d) the function $\alpha_f(s) = \limsup_{z\to o} \frac{[f(s,0)-f(s,z)]^+}{|z|}$ belongs to $L^1_{loc}(\mathbb{R})$, where $[t]^+ = \max\{0,t\}$.
>
> Then we have :

i) for every $u \in W^{1,1}_{loc}(\Omega)$ the function $x \to f(u(x),Du(x))$ is measurable on Ω ;

ii) the functional $F(u) = \int_\Omega f(u,Du)dx$ is lower semicontinuous on $W^{1,1}_{loc}(\Omega)$ with respect to the topology induced by $L^1_{loc}(\Omega)$.

An example of a functional to which we can apply theorem 1 is the following.

Example 1

The hypotheses of theorem 1 are fulfilled by functionals of the form

$$F(u) = \int_\Omega \left[\sum_{i,j=1}^{n} a_{ij}(u)\, D_i u\, D_j u + \sum_{i=1}^{n} b_i(u) D_i u + c(u) \right] dx$$

where a_{ij}, b_i are measurable functions, $b_i \in L^1_{loc}(\mathbb{R})$, c is lower semicontinuous and

$$\sum_{i,j=1}^{n} a_{ij}(s) z_i z_j + \sum_{i=1}^{n} b_i(s) z_i + c(s) \geqslant 0 \quad \text{for every } s \in \mathbb{R},\ s \in \mathbb{R}^N .$$

Hypothesis (d) in theorem 1 cannot be dropped as the following example shows.

Example 2 (see [9]).

Let $n=1$, $\Omega =]0,1[$ and let $f(s,z)$ be defined by

$$f(s,z) = \begin{cases} (1 + \frac{z}{s})^+ & \text{if } d \neq 0 \\ 1 & \text{if } s = 0 . \end{cases}$$

For every $\varepsilon > 0$ let $u_\varepsilon(x) = \varepsilon - \varepsilon x$. Then u_ε converges to 0 as $\varepsilon \to 0$, but $F(u_\varepsilon) = 0$, whereas $F(0) = 1$. Note that f satisfies conditions (a), (b), (c) of theorem 1 .

We give now a sketch of the proof of theorem 1.

Definition 1

We say that a function $g : \mathbb{R} \times \mathbb{R}^n \to \mathbb{R}$ is an admissible integrand if

i) for every $z \in \mathbb{R}^n$ the function $g(\cdot,z)$ is measurable ;

ii) for a.a. $s \in \mathbb{R}$ the function $g(s,\cdot)$ is continuous ;

iii) the function $g(\cdot,0)$ is a Borel function.

Moreover we say that two admissible integrands g_1, g_2 are equivalent if

iv) for every $s \in \mathbb{R}$ $g_1(s,0) = g_2(s,0)$;

v) there exists a Borel set $N \subseteq \mathbb{R}$, with $\text{meas}(N) = 0$, such that

$$g_1(s,z) = g_2(s,z) \quad \text{for every} \quad s \in \mathbb{R} - N \text{ and } z \in \mathbb{R}^n .$$

Proposition 2 (see [9]).

If g_1, g_2 are equivalent integrands, then for every $u \in W^{1,1}_{loc}(\Omega)$

$$g_1(u(x),Du(x)) = g_2(u(x),Du(x)) \quad \text{a.e. on } \Omega .$$

Assertion (i) of theorem 1 follows now from proposition 2 and from the fact that, for every admissible integrand g_1, there exists an equivalent integrand g_2 such that g_2 is a Borel function from $\mathbb{R}^{n+1}$ into $\mathbb{R}$. The essential steps to prove assertion (ii) of theorem 1 are the following.

Step 1

We prove the theorem when

$$f(s,z) = [\, a(s) + \langle b(s),z \rangle]^+$$

with a,b measurable functions, $a \leqslant 0$.

Step 2

We next prove the following lemma :

LEMMA 1 (see [9]).

Let (f_h) be a sequence of non-negative admissible integrands and let $f_\infty = \sup_h f_h$. Set for every open subset A of Ω , every $u \in W^{1,1}_{loc}(A)$ and every $h \in \mathbb{N} \cup \{\infty\}$

$$F_h(u,A) = \int_A f_h(u,Du)dx .$$

Suppose that for every $h \in \mathbb{N}$ and every open subset A of Ω the functional $F_h(\cdot,A)$ is $L^1_{loc}(A)$-lower semicontinuous on $W^{1,1}_{loc}(A)$. Then, for every open subset A of Ω the functional $F_\infty(\cdot,A)$ is $L^1_{loc}(A)$-lower semicontinuous on $W^{1,1}_{loc}(A)$.

Step 3

In the case $f(s,0) = 0$ the proof of assertion (ii) of theorem 1 follows from the previous steps by using the fact that, since $f(s,z)$ is convex in z, we have

$$f(s,z) = \sup_{h \in \mathbb{N}} [a_h(s) + \langle b_h(s), z \rangle]^+$$

for suitable measurable functions a_h, b_h with $a_h \leq 0$.

Step 4

From the case $f(s,0) = 0$ we can pass to the general case by using hypothesis (d).

Theorem 1 suggests many problems : the first one is that of finding sufficient conditions on the function f such that for every $u \in W^{1,1}_{loc}(\Omega)$

$$(1) \qquad \Gamma(L^1_{loc}(\Omega)^-) \int_\Omega f(u,Du)dx = \int_\Omega \varphi(u,Du)dx$$

for a suitable integrand φ.

If we want to seek a characterization of φ, it is useful to introduce an ordering consistent with proposition 2 and to set for every pair f_1, f_2 of functions from $\mathbb{R} \times \mathbb{R}^n$ into $\mathbb{R}$

$$f_1 \precsim f_2 \iff \begin{cases} \text{i)} \ f_1(s,0) \leq f_2(s,0) & \text{for every } s \in \mathbb{R}; \\ \text{ii)} \ \text{there exists a Borel subset } N \subseteq \mathbb{R}, \text{ with meas}(N) = 0, \\ \quad \text{such that } f_1(s,z) \leq f_2(s,z) \ \text{ for every } s \in \mathbb{R}-N \text{ and } z \in \mathbb{R}^n. \end{cases}$$

It is still an open problem to establish under what hypotheses on f the function φ in (1) is the greatest function less than or equal to f with respect to the ordering $\precsim$, which statisfies conditions (a), (b), (c) of theorem 1. An example of function f for which this happens is

$$f(s,z) = |z|^2 + \alpha(s)$$

where

$$\alpha(s) = \begin{cases} 0 & \text{if } s \text{ is rational} \\ 1 & \text{if } s \text{ is irrational.} \end{cases}$$

In this case it possible to calculate explicitly the function φ which is given by

$$\varphi(z) = \begin{cases} 1 + |z|^2 & \text{if } |z| \geq 1 \\ 2|z| & \text{if } |z| < 1. \end{cases}$$

It is easy to see that, without additional hypotheses, theorem 1 cannot be extended to functionals of the form

$$(2) \qquad F(u) = \int_\Omega f(x,u,Du)\,dx$$

where $f(x,s,z)$ is measurable with respect to x, and where u can be a vector-valued function.

It would be very interesting to find lower-semicontinuity results which contain both theorem 1 and the results established by other authors in the case of functionals of type (2) with $f(x,s,z)$ continuous or lower-continuous with respect to s.

With respect to the relaxation problems, I want to point out two conjectures.

<u>Conjecture 1</u> (Scalar case)

Let $f : \Omega \times \mathbb{R} \times \mathbb{R}^n \to \mathbb{R}$ be a Borel function such that

$$0 \leqslant f(x,s,z) \leqslant c(1 + |s|^\alpha + |z|^\alpha) \qquad (\alpha \geqslant 1)$$

for every (x,s,z). Then, for a suitable function φ, we have

$$\Gamma(L^\alpha(\Omega)^-) \int_\Omega f(x,u,Du)\,dx = \int_\Omega \varphi(x,u,Du)\,dx$$

for every $u \in W^{1,\alpha}(\Omega)$.

<u>Conjecture 2</u> (Vector case)

Let $f : \Omega \times \mathbb{R}^m \times \mathbb{R}^{nm} \to \mathbb{R}$ be a Borel function such that

$$|z|^\alpha \leqslant f(x,s,z) \leqslant c(1 + |s|^\alpha + |z|^\alpha) \qquad (\alpha \geqslant 1)$$

for every (x, s,z). Then, for a suitable function φ, we have

$$\Gamma([L^\alpha(\Omega)^m]^-) \int_\Omega f(x,u,Du)\,dx = \int_\Omega \varphi(x,u,Du)\,dx$$

for every $u \in [W^{1,\alpha}(\Omega)]^m$.

A good reason to think that the answer of Conjectures 1 and 2 is positive is that, in both cases, for every $u \in W^{1,\alpha}(\Omega)$, the set function $A \to \Gamma(L^\alpha(A)^-)$ $\int_A f(x,u,Du)\,dx$ defined on the open subsets of Ω is the restriction of a regular Borel measure (see [4]).

It would be very interesting to extend our previous remarks about integrals depending on first derivatives to integrals depending on higher order derivatives. For example consider the functional

$$F(u) = \int_\Omega [|\Delta u|^2 + \alpha(u)]dx$$

where Δ denotes the Laplace operator and

$$\alpha(s) = \begin{cases} 0 & \text{if } s \text{ is rational} \\ 1 & \text{if } s \text{ is irrational.} \end{cases}$$

Then it seems natural to put forth the following

Conjecture 3

For every $u \in W^{2,2}(\Omega)$ we have

$$\Gamma(L^1_{loc}(\Omega)^-)\ F(u) = \int_\Omega [|\Delta u|^2 + \beta(Du)]dx$$

where

$$\beta(z) = \begin{cases} 0 & \text{if } z = 0 \\ 1 & \text{if } z \neq 0 . \end{cases}$$

In addition to functionals of the form $\int_\Omega f(x,u,Du)dx$, in many problems in the Calculus of Variations there arise integrals such as

(3) $$\int_\Omega f(x,u,Du)dx + \int_{\overline{\Omega}} g(x,u)\ d\mu(x)$$

where μ is a measure on $\overline{\Omega}$.

For example, if Ω is a smooth bounded open set and $u \in C^2(\Omega) \cap C(\overline{\Omega})$ is a solution of the problem

$$\begin{cases} \operatorname{div}\left(\dfrac{Du}{\sqrt{1+|Du|^2}}\right) = 0 & \text{in } \Omega \\ u = \varphi & \text{on } \partial\Omega , \end{cases}$$

then u minimizes the area integral

$$\int_\Omega \sqrt{1+|Du|^2}dx + \int_{\partial\Omega} |u - \varphi|dH^{n-1}$$

where H^{n-1} is the $n-1$ dimensional Hausdorff measure.

If on the contrary $u \in C^2(\Omega) \cap C(\overline{\Omega})$ is a solution of the problem

$$\begin{cases} \Delta u = u + g(x) & \text{in } \Omega \\ \dfrac{\partial u}{\partial \nu} = f & \text{on } \partial\Omega \end{cases}$$

(ν denoting the inner normal vector on $\partial\Omega$), then u minimizes the integral

$$\int_\Omega [|Du|^2 + |u|^2 + 2ug(x)]dx + 2\int_{\partial\Omega} u\, f(x)\, dH^{n-1} ;$$

In the theory of variational inequalities and in minimum problems with obstacles, functionals of the type (3) often occur with μ having support not contained in $\partial\Omega$ (see for example [3], [5], [7], [10]). To my knowledge, however, general semicontinuity and relaxation theorems for functionals of the type (3) fail. To begin with, we may consider the case when f, g are continuous non-negative functions, $a \geq 1$, $\int_\Omega f(x,u,Du)dx$ is $L^1_{loc}(\Omega)$--lower semicontinuous on $W^{1,1}_{loc}(\Omega)$. Then the following problems arise.

<u>Problem 3</u>

Find conditions under which the functional

$$\int_\Omega f(x,u,Du)dx + \int_{\overline{\Omega}} g(x,u)\, d\mu(x)$$

is $L^\alpha(\Omega)$-lower semicontinuous on $W^{1,1}(\Omega) \cap C(\overline{\Omega})$.

<u>Problem 4</u>

If $\int_\Omega f(x,u,Du)dx + \int_{\overline{\Omega}} g(x,u)\, d\mu(x)$ is not $L^\alpha(\Omega)$-lower semicontinuous, then we may ask if there exists a function $\gamma(x,s)$ such that for every $u \in W^{1,1}(\Omega) \cap C(\overline{\Omega})$

$$(4) \quad \Gamma(L^\alpha(\Omega)^-)\,[\int_\Omega f(x,u,Du)dx + \int_{\overline{\Omega}} g(x,u)\, d\mu(x)] = \int_\Omega f(x,u,Du)dx + \int_{\overline{\Omega}} \gamma(x,u)\, d\mu(x) .$$

<u>Problem 5</u>

Suppose there exists a function γ satisfying (4) and set for every $u \in L^\alpha(\Omega)$

$$F(u) = \Gamma(L^\alpha(\Omega)^-) \int_\Omega f(x,u,Du)dx$$

$$\Phi(u) = \Gamma(L^\alpha(\Omega)^-)\, [\int_\Omega f(x,u,Du)dx + \int_{\overline{\Omega}} g(x,u)\, d\mu(x)] .$$

Then, one may ask if for every $v \in L^\alpha(\Omega)$ with $F(v) < +\infty$, and for every $x \in \overline{\Omega}$, there exists a set $Tr(x,v,f,\mu)$ such that

$$\Phi(v) = F(v) + \int_{\overline{\Omega}} \inf\{\gamma(x,t) : t \in Tr(x,v,f,\mu)\} d\mu(x).$$

Problem 5 introduces the concept of "variational trace" Tr ; I believe it contains many instances of traces which usually occur in Calculus of

Variations and in Partial Differential Equations.

For example, if Ω is an open subset of $\mathbb{R}^2$, C is a closed regular curve in Ω, $\mu(B) = 2H^1(C \cap B)$ for every Borel subset $B \subseteq \Omega$, $f(x,s,z) = \sqrt{1+|z|^2}$, then

$$\gamma(x,t) = \min\{ g(x,s) + |s-t| : s \in \mathbb{R} \},$$

the condition $F(v) < +\infty$ is equivalent to $v \in BV(\Omega)$, and $Tr(x,v,f,\mu)$ is the set of $t \in \mathbb{R}$ such that

$$\lim_{\rho \to o^+} ⨍_{B_\rho(x)} |v(y) - t|\,dy = \lim_{\rho \to o^+} \min_{\tau \in \mathbb{R}} ⨍_{B_\rho(x)} |v(y) - \tau|\,dy$$

where $B_\rho(x) = \{ y \in \mathbb{R}^n = |x-y| < \rho \}$ and $⨍_A f(y)\,dy$ denotes the average of f on A.

Finally, I should like to point out that another interesting class of functionals is the one consisting of integrals of the form

$$(5) \quad \int_\Omega \{ f(x,u,v,Du,Dv,\ldots,D^m u,D^m v) + J[g(x,u,v,Du,Dv,\ldots,D^m u,D^m v)] \}\,dx$$

where f,g are sufficiently regular functions and J is defined by

$$J[t] = \begin{cases} 0 & \text{if} \quad t = 0 \\ +\infty & \text{if} \quad t \neq 0 . \end{cases}$$

Many problems in Optimal Control Theory can be written as problems of Calculus of Variations for functionals of the type (5) ; therefore, it would be very useful to try to construct a general theory of semicontinuity and relaxation which permits us to treat at the same time functionals of type (3) and functionals of type (5).

ACKNOWLEGDMENTS

I wish to thank Dr. G. BUTTAZZO for his most useful collaboration in the preparation of this paper.

REFERENCES

[1] E. ACERBI, G. BUTTAZZO, N. FUSCO : Semicontinuity and relaxation for integrals depending on vector-valued functions. J. Math. Pures Appl., 62 (1983), 371 - 387.

[2] E. ACERBI, N. FUSCO : Semicontinuity Problems in Calculus of Variations ; Archive for Rat. Mech. and Anal. (to appear).

[3] H. ATTOUCH, C. PICARD : Variational inequalities with varying obstacles : the general form of the limit problem. J. Funct. Anal., 50 (1983), 329-386.

[4] G. BUTTAZZO, G. DAL MASO : Γ- limits of integral functionnals ; J. Analyse Math 37 (1980), 145-185.

[5] L. CARBONE, F. COLOMBINI : On convergence of functionals with unilateral constraints ; J. MAth. Pures Appl. 59 (1980), 465-500.

[6] L. CESARI : Lower semicontinuity and lower closure theorems without seminormality conditions. Ann. Mat. Pura Appl., 98 (1974), 382-397.

[7] G. DAL MASO, P. LONGO : Γ-limits of obstacles ; Ann. Mat. Pura Appl. (4) 128 (1980), 1-50.

[8] E. DE GIORGI : Teoremi di semicontinuità nel Calcolo delle Variazioni. Seminario, Istituto Nazionale di Alta Matematica, Roma 1968-1969.

[9] E. DE GIORGI, G. BUTTAZZO, G. DAL MASO : On the lower semicontinuity of certain integral functionals. Att Accad. Naz. Lincei, (to appear).

[10] E. DE GIORGI, G. DAL MASO, P. LONGO : Γ-limiti di ostacoli. Atti Accad. Naz. Lincei. Rend. Cl. Sci. Fis. Mat. Natur., (8) 68 (1980), 481-487.

[11] I. EKELAND, R. TEMAM : Convex analysis and variational problems ; North-Holland, Amsterdam (1978).

[12] P. MARCELLINI, C. SBORDONE : Semicontinuity problems in the Calculus of Variations. Nonlinear Anal 4 (1980), 241-257.

[13] C.B. MORREY : Multiple integrals in Calculus of Variations. Springer-Verlag, Berlin (1966).

[14] C. OLECH : Weak lower semicontinuity of integral functionals. J. Optimization Theory Appl., 19 (1976), 3.16.

[15] J. SERRIN : On the definition and properties of certain variational integrals. Trans. Amer. Math. Soc., 101 (1961), 139-167.

Ennio DE GIORGI
Scula Normale Superiore
Piazza dei Cavalieri, 7
I - 56100 - PISA
Italy

J P AUBIN

Variational problems related to the Monge–Ampère equation

1°) The Functional

Let Ω be a bounded strictly convex set of $\mathbb{R}^n$. For a strictly convex function $\varphi \in C^2(\overline{\Omega})$ vanishing on $\partial\Omega$, we define

$$I(\varphi) = -n \int_\Omega \varphi \det((\partial_{ij}\varphi))dx$$

$\{x^i\}$ being the coordinates of $\mathbb{R}^n$, $\partial_{ij}\varphi$ means $\partial^2\varphi/\partial x^i \partial x^j$ and we set $M(\varphi) = \det((\partial_{ij}\varphi))$. Let us consider the Riemannian metric $g_{ij} = \partial_{ij}\varphi$. We verify that

$$I(\varphi) = \int_\Omega g^{ij} \partial_i\varphi \; \partial_j\varphi \; M(\varphi)dx \; ;$$

indeed, $n\, M(\varphi) = \partial_i [g^{ij} M(\varphi) \partial_j \varphi]$ and integrating by parts gives the result.

If the function is radially symmetric $\varphi(x) = f(\|x\|)$, the integrand is $g^{ij} \partial_i\varphi \partial_j\varphi M(\varphi)dx = f'^{n+1}(r)\, dr\, d\omega$ with $r = \|x\|$ and $\omega \in S_{n-1}(1)$. For a radially symmetric function $\varphi \in C^1(\overline{B}_\tau)$ which is zero on $\partial\overline{B}_\tau$, we define, if $f' \geqslant 0$, $I(\varphi) = \sigma_{n-1} \int_o^\tau f'^{n+1}(r)\, dr$. Here B_τ is the ball in $\mathbb{R}^n$ of radius τ and σ_{n-1} is the volume of the sphere $S_{n-1}(1)$. The definition of $I(\varphi)$ extends to any convex function vanishing on the boundary. We proceed as follows. For a convex function φ on Ω it is possible to define a Radon measure $\mathcal{M}(\varphi)$ which is equal to $\mathcal{M}(\varphi) = M(\varphi)\, dx$ when $\varphi \in C^2$. We then set

$$I(\varphi) = -n \lim_{\substack{< \\ a \to o}} \int_{\Omega_a} \varphi \mathcal{M}(\varphi) \quad \text{with} \quad \Omega_a = \{x \in \Omega / \varphi(x) \leqslant a < o\}.$$

We define $\mathcal{M}(\varphi)$ by induction on the dimension. Let us show how to begin. Let $J \subset \mathbb{R}$ be a bounded open interval, let φ be a convex function on J and let $\psi \in C_o(J)$ be a continuous function on J with support $K \subset J$.

There exists $h \in \mathscr{D}(J)$, with support $\tilde{K} \subset J$, $h \geqslant 0$ and $h/K = 1$. If $\varphi \in C^2$ we can write

$$\left| \int_J \psi \, \partial_{11} \varphi \, dt \right| \leqslant \sup |\psi| \int_J h \, \partial_{11} \varphi \, dt = \sup |\psi| \int_J \varphi \, \partial_{11} h \, dt \leqslant$$

$$\text{Cons.} \sup |\psi| \sup_{\tilde{K}} |\varphi| .$$

If φ is only convex, let $\{\varphi_i\}$ be a sequence of C^∞ convex functions such that $\varphi_i \to \varphi$ uniformly on $\tilde{K}$. $u_i = \int_J \psi \partial_{11} \varphi_i \, dt$ is a Cauchy sequence : $|u_i - u_j| \leqslant$ Const. $\sup |\psi| \sup_{\tilde{K}} |\varphi_i - \varphi_j|$ and we define

$\int \psi \mathscr{M}(\varphi) = \lim_{i \to \infty} u_i$. This definition is valid and it is obvious that the limit does not depend on the sequence $\{u_i\}$.

2°) <u>The variational Problem</u>

Let $f(u) \in C^k\,(]-\infty, 0\,])$ be a strictly positive function when $u \neq 0$, and $f(u) > \varepsilon > 0$ for $u < -1$ $(k \geqslant 0)$. We set

$F(t) = \int_{-|t|}^{0} f(u)\, du$ and $\mathscr{A}(\varphi) = \int_B F(\varphi(x))\, dx$ for all continuous functions on the unit ball $B \subset \mathbb{R}^n$.

Choose $C > 0$ and define $\mu = \inf I(\varphi)$ for all convex functions φ in B vanishing on ∂B and satisfying $\mathscr{A}(\varphi) = C$.

<u>Theorem 1</u>

μ is attained by a radially symmetric convex function $\varphi_o \in C^{k+2}(\bar{B})$, $\varphi_{o/\partial B} = 0$, and φ_o satisfies, for some $a > 0$, the Monge-Ampère Equation :

$$\det((\partial_{ij} \varphi_o)) = a\, f(\varphi_o) .$$

Making use of § 3 reduces the problem to a one-dimensional one. With this result it is possible to solve equations having a more general right-hand side, by using the method of upper and lower solutions.

3°) <u>Symmetrization procedure</u>

Let Ω be a bounded convex set of $\mathbb{R}^n$. Pick $P \in \Omega$ and let H be the projection of P on the tangent plane to $\partial\Omega$ at a point $Q \in \partial\Omega$.

ω will denote the oriented direction of $\vec{PH}$ ($\omega \in S_{n-1}(1)$) and $\ell(\omega) = \overline{PH}$, the length of PH. Let d(resp. D) be the inradius (resp. the circumradius) of Ω.

If Θ is another bounded convex set of $\mathbb{R}^n$, write

$$g(\rho) = \mathrm{vol}^{1/n} [(1-\rho)\Omega + \rho\Theta] \text{ for } \rho \in [0,1].$$

Using the fundamental result of Minkowski : $g(\rho)$ is a concave function in ρ, we find Minkowski's inequalities. In particular, the first mixed volume $V_1(\Omega,\Theta)$ satisfies $V_1^n(\Omega,\Theta) \geqslant \mu(\Omega)\ [\mu(\Theta)]^{n-1}$ where $\mu(\Theta)$ is the volume of Θ. By definition, $V_1(\Omega,\Theta) = \frac{1}{n}\int_{S_{n-1}} \ell_\Omega(\omega)\, d\sigma_\Theta(\omega)$, $d\sigma_\Theta$ being the area element of $\partial\Theta$. We have also

$$n V_1(\Omega,\Theta) = \lim_{h\to 0} \frac{1}{h} [\mu(\Theta + h\Omega) - \mu(\Theta)].$$

If $\Omega = B$, we obtain the usual isoperimetric inequality.

If $\Theta = B$, we find the following inequality useful later :

(1) $$\left(\frac{1}{n}\int_{S_{n-1}} \ell_\Omega(\omega)\, d\omega\right)^n \geqslant \mu(\Omega)\ [\mu(B)^{n-1}].$$

<u>Proposition 2</u>

Let $\varphi \in C^2(\bar{\Omega})$ be a strictly convex function in Ω which is zero on $\partial\Omega$. There exists a radially symmetric function $\tilde{\varphi} \in C^1(\bar{B}_\tau)$ with $d \leqslant \tau \leqslant D$, $\tilde{\varphi}$ being zero on $\partial\bar{B}_\tau$ and satisfying the following properties.

α) $\tilde{\varphi}$ and φ have the same extrema

β) $I(\tilde{\varphi}) \leqslant I(\varphi)$

γ) $\mu(\Omega_a) \leqslant \mu(\tilde{\Omega}_a)$ for all $a < 0$ where $\tilde{\Omega}_a = \{x \in B_\tau / \tilde{\varphi}(x) \leqslant a\}$.

<u>Proof</u>

Define $\Sigma_a = \{x \in \Omega / \varphi(x) = a\}$ and $m = \inf \varphi$. Integrating on Σ_a yields

$$I(\varphi) = \int_m^0 da \int_{\Sigma_a} \frac{|\nabla\varphi|^n}{\prod_{i=1}^{n-1} R_i}\, d\sigma$$

R_i being the principal radii of curvature of Σ_a.

But as $d\sigma = \prod_{i=1}^{n-1} R_i\, d\omega$, in fact we have

$$I(\varphi) = \int_m^0 da \int_{\Sigma_a} |\nabla\varphi|^n\, d\omega$$

On the other hand, using Hölder's inequality yields

$$(2) \qquad \sigma_{n-1} = \int_{\Sigma_a} d\omega \leqslant \left(\int_{\Sigma_a} |\nabla\varphi|^n d\omega\right)^{1/(n+1)} \left(\int_{\Sigma_a} \frac{d\omega}{|\nabla\varphi|}\right)^{n/(n+1)} .$$

Set $r(a) = \frac{1}{\sigma_{n-1}} \int_{\Sigma_a} \ell(\omega)\, d\omega$. $r(a)$ is a strictly increasing function which is C^1 on $]m,0[$, and $r'(a) = \frac{1}{\sigma_{n-1}} \int_{\Sigma_a} \frac{d\omega}{|\nabla\varphi|}$.

Let $r \to g(r) = a$ be the inverse function and set $\tilde{\varphi}(x) = g(\|x\|)$.

Let us verify the conclusions of proposition 2 : α) is obvious.

As equality (2) holds for $\tilde{\varphi}$, $\int_{\partial\tilde{\Omega}_a} |\Delta\tilde{\varphi}|^n\, d\omega \leqslant \int_{\Sigma_a} |\nabla\varphi|^n\, d\omega$ for all $a < 0$; integrating over $[m,0]$ gives β). To prove γ) we use inequality (1) : among the convex bodies for which $\int_{\partial\Omega} \ell(\omega)\, d\omega$ is given, the ball has the greatest volume. By construction it is the case of Ω_a and $\tilde{\Omega}_a$, thus $\mu(\Omega_a) \leqslant \mu(\tilde{\Omega}_a)$.

4°) Proof of Theorem 1

First of all we prove that μ is a equal to the inf. of $I(\varphi)$ when we consider only convex functions $\varphi \in C^2(\bar{B})$. Then let $\varphi \in C^2(\bar{B})$ be a convex function which is zero on ∂B , and let $\tilde{\varphi}$ be the corresponding radially symmetric function defined above. As $F(t)$ is an increasing function in $|t|$, $\gamma) \Rightarrow \mathscr{A}(\varphi) \leqslant \mathscr{A}(\tilde{\varphi})$. Then there exists a function $\psi = \xi\tilde{\varphi}$ with $\xi \leqslant 1$ such that $\mathscr{A}(\psi) = C$ and $I(\psi) \leqslant I(\tilde{\varphi}) \leqslant I(\varphi)$. Therefore μ is equal to the inf. of $I(\psi)$ for all radially symmetric functions $\psi \in C^1(\bar{B})$ $(\psi(x) = g\|x\|)$ satisfying $\psi/_{\partial B} = 0$, $g' \geqslant 0$ and $A(\psi) = C$.

5°) Solving of the variational problem in one dimension

We consider the inf. of $I(g)$ when $g \in H_1^{n+1}([0,1])$ satisfies

$$\mathscr{A}(g) = C \ , \quad g \leqslant 0 \ \text{and} \quad g(1) = 0 \ , \text{ with } \ I(g) = \sigma_m \int_0^1 |g'(r)|^{n+1}\, dr.$$

The sketch of the proof is the following. Let $\{g_i\}$ be a minimising sequence. These functions are equicontinuous :

$$|g_i(b) - g_i(a)| = \left|\int_a^b g'(r)dr\right| \leqslant \left|\int_a^b |g'(r)|^{n+1} dr\right|^{1/(n+1)} |b-a|^{n/(n+1)} .$$

By Ascoli's theorem, there exist a continuous function g_o on $[0,1]$ and $\{g_j\}$ a subsequence of $\{g_i\}$ such that $g_j \to g_o$ uniformly. Thus $\mathscr{A}(g_o) = C$, $g_o \leqslant 0$ and $g_o(1) = 0$. Moreover a subsequence $\{g_k\}$ converges weakly to g_o in $H_1^{n+1}([0,1])$. Therefore the inf. of $I(g)$ is attained at g_o and the Euler equation is :

$$\int_o^1 \psi' |g_o'|^{n-1} g_o' dr = -\nu \int_o^1 \psi f(g_o) r^{n-1} dr$$

for all functions $\psi \in H_1^{n+1}([0,1])$ which are zero at $r = 1$, ν being a constant. It is easy to verify that $g_o(r)$ is equal to

$$\tilde{g}(r) = \int_1^r \left[\nu \int_o^u f[g_o(t)] t^{n-1} dt \right]^{1/n} du .$$

Bibliography

[1] AUBIN T. Nonlinear Analysis on Manifolds. Monge-Ampère Equations Springer (1982).

[2] BUSEMAN H. Convex surfaces. Interscience Tracts in Pure and Appl. Math. 6 (1958).

[3] RAUCH J. and TAYLOR B. The Dirichlet Problem for the multidimensional Monge-Ampère Aquation. Rocky Mountain J. of Math 7 (1977), 345-363 .

Thierry AUBIN
Département de Mathématiques
et Laboratoire LA 213
Université de PARIS VI
4, Place Jussieu
75005 - PARIS

L BOUTET DE MONVEL

The index theorem for almost elliptic systems

This is an account of a joint work with B. MALGRANGE. We use an extension of the notion of almost ellipticity, which allows embedding in the algebraic sense to give a natural proof of the index theorem. The proof is otherwise modelled on Grothendieck's proof of the Riemann-Roch [5] theorem: everything is embedded in a numeric space, where the theorem is known (this method is also the starting point of the printed proof of Atiyah and Singer [3] but it is technically a little harder to embed within the category of elliptic systems).

§ 1 - Almost Elliptic Systems

Let first Ω be a complex manifold with smooth real boundary $\partial\Omega$, and let D be a complex of differential operators on Ω :

$$(1) \qquad 0 \longrightarrow \mathcal{O}(\Omega,E_o) \xrightarrow{D_o} \mathcal{O}(\Omega,E_1) \xrightarrow{D_1} \dots \mathcal{O}(\Omega,E_n) \longrightarrow 0$$

i.e. the E_j's are holomorphic vector bundles, $\mathcal{O}$ denotes the space of holomorphic sections, the D_j's are holomorphic differential operators such that $D_{j+1} \circ D_j = 0$.

The symbol $\sigma(D)$ is a holomorphic complex of vector bundles on the cotangent bundle $T^\star\Omega$:

$$(2) \qquad 0 \longrightarrow p^\star E_o \xrightarrow{d_o} p^\star E_1 \xrightarrow{d_1} \dots \; p^\star E_n \longrightarrow 0$$

where $p^\star E_j$ is the pull-back of E_j, and d_j is determined by the fact that for any holomorphic function φ and section u we have

$$e^{-t\varphi} D_j(e^{t\varphi} u) = t^m d_j(d\varphi).u + O(t^{m-1})$$

if m is the order of D_j.

<u>Definition 1</u> - <u>The differential complex D is elliptic on Ω if $\sigma(D)$ is exact at all normal covectors on the boundary</u>

i.e. $\sigma(D)(\partial\rho)$ is exact if $\rho = 0$ is a real defining equation for $\partial\Omega$, and $\partial\rho$ denotes the holomorphic part of the derivative of ρ.

Let now X be a compact real analytic manifold with analytic boundary ∂X. Let $\tilde{X}$ be a complexification of X, i.e. $\tilde{X}$ is a complex manifold of dimension $\dim_{\mathbb{C}} \tilde{X} = \dim_{\mathbb{R}} X$, with a closed totally real submanifold $\mathrm{Re}\tilde{X}$ in which X is embedded (such a manifold always exists).

Le ρ be a real analytic function on $\tilde{X}$, such that $\rho < 0$ on $X - \partial X$, $\rho = 0$ and $\partial\rho \neq 0$ on ∂X. Let u be a real analytic function on $\tilde{X}$ such that $u > 0$ outside of $\mathrm{Re}\tilde{X}$, and u vanishes transversally elliptically on $\mathrm{Re}\tilde{X}$. Such functions exist if $\tilde{X}$ is small enough ; for example if X is the unit ball of $\mathbb{R}^n$, we may choose $\tilde{X} = \mathbb{C}^n$, $\rho = |x|^2 - 1$, $u = |y|^2$ with $z = x + iy \in \mathbb{C}^n$.

We denote X_ε the set

(3) $$X_\varepsilon = \{ z \in \tilde{X} , u(z) + \varepsilon\rho(z) \leqslant \varepsilon^2 \} .$$

If $\tilde{X}$ and $\varepsilon > 0$ are small enough, X_ε is a compact, strictly pseudo-convex manifold with real analytic boundary, and the X_ε form a fundamental system of neighbourhoods of X in $\tilde{X}$.

Let D be a complex of analytic differential operators on X. Then D extends to X_ε for small $\varepsilon > 0$.

<u>Definition 2</u> - <u>We will say that D is almost elliptic if its extension to X_ε is elliptic (def. 1) for all small $\varepsilon > 0$</u>.

<u>Example 1</u> - If X has no boundary and D is elliptic in the usual sense (i.e. $\sigma(D)$ is exact at real non zero covectors), then it is almost elliptic : in this case we may take $\rho = -1$, the normal covector on ∂X_ε is ∂u, and for small $\varepsilon > 0$ the direction of ∂u is asymptotic to a pure imaginary direction.

<u>Example 2</u> - Let $\mathrm{Char}(D)$ be the set of all complex covectors on $\tilde{X}$ where $\sigma(D)$ is not exact (it is an analytic cone, of dimension $\geqslant \dim X$). Then D is almost elliptic if it is elliptic on X in the following sense :

i) D is elliptic on $\mathrm{Re}\tilde{X}$ (near X).

ii) $\mathrm{Char}(D)$ contains no complex covector above ∂X of the form $\partial\rho + i\xi$, with ξ real.

The second condition means that the Shapiro - Lopatinsky boundary ellipticity condition is fulfilled without adding boundary conditions. Equivalently, at every boundary point, the tangent homogeneous system with constant coefficients on a half space has no non-constant exponential solution that is bounded on the boundary.

Example 3 - Let D be a complex of differential operators on a complex manifold with boundary Ω . Let Ω_R be the real manifold defined by Ω ; its complexification contains a neighbourhood of the diagonal in $\Omega \times \bar{\Omega}$. Let $\bar{\partial}$ denote the Cauchy-Riemann complex of Ω_R , i.e. the De Rham complex of $\bar{\Omega}$. Then D is essentially equivalent to the complex $D_R = D \otimes \bar{\partial}$ on Ω_R . One checks that D is elliptic as in def. 1 if and only if D_R is elliptic as in example 2 .

Proposition 1 - If D is holonomic, it is almost elliptic.

Holonomic means that $\Lambda = \text{char}(D)$ is Lagrangian, i.e. since it is conic, that the Liouville form $\lambda = \Sigma \xi_j \frac{\partial}{\partial x_j}$ (in local coordinates) vanishes on it. With notation as above, the ∂X_ε are the level surfaces of $\varphi = \rho + \sqrt{\rho^2 + 4u}$. Let $N \subset T^\star \tilde{X}$ be the submanifold image of the section $\partial\varphi$. Let Φ be the pull back of φ to $T^\star \tilde{X}$. Then on $N \cap \Lambda$ we have $\partial\Phi = \lambda = 0$. Since Φ is real, we also have $\bar{\partial}\Phi = 0$ so Φ is locally constant on $N \cap \Lambda$. Since every thing is real analytic and X is compact, 0 is an isolated value of Φ on $N \cap \Lambda$ (near X), which proves the proposition.

Proposition 2 - Let X_1 and X_2 be two real analytic manifolds with $\partial X_2 = \emptyset$. If D_1 and D_2 are complexes of differential operators on X_1 and X_2, the product complex $D_1 \otimes D_2$ on $X_1 \times X_2$ is well defined. It is almost elliptic if D_1 and D_2 are almost elliptic.

Proof : let ρ_1 , u_1 , u_2 be as above for X_1 and X_2 . On $\tilde{X} = \tilde{X}_1 \times \tilde{X}_2$ we choose $\rho = \rho_1$, $u = u_1 + u_2$. Any normal covector on ∂X_ε is proportional to $\partial\rho$. At such a point we have $\sigma(D_1) = \sigma(D_1)\ (\partial(\varepsilon\rho_1 + u_1))$, $\sigma(D_2) = \sigma(D_2)\ (\partial u_2)$. If ε is small enough and > 0 , one of the two numbers $\varepsilon\rho_1 + u_1$ and u_2 is > 0 (and both are small) ; the corresponding covector $\partial(\varepsilon\rho_1 + u_1)$, ∂u_2 does not vanish, so one of the two $\sigma(D_1)$, $\sigma(D_2)$ is exact, and so is the product $\sigma(D_1) \otimes \sigma(D_2)$.

Example 4 - Let V be a vector space, and let k_v be the Koszul complex, viewed as a complex of differential operators

$$(4) \qquad 0 \to O(V, \Lambda^n V^\star) \xrightarrow{k} O(V, \Lambda^{n-1} V^\star) \xrightarrow{k} \dots O(V, V^\star) \to 0 \to 0 .$$

The differential K at $y \in V$ is the interior product $\omega \to y_L \omega$. Here our complex is indexed by negative integers, i.e. the j-th vector boundle is $E_j = \Lambda^{-j} V^\star$.

The Koszul complex is exact oustide of the origin, hence holonomic, and almost elliptic. So if D is almost elliptic on X , $D \otimes K$ is almost elliptic on X x V .

§ 2 - Description of the index theorem

a. Index. Let D be a complex of differential operator on a real or complex manifold X . It is known that if D is elliptic then the homology groups $H^j(D) = \text{Ker } D_j / \text{Im } D_{j-1}$ are finite dimensional, so one can define the index (Euler characteristic) :

$$\text{Ind}(D) = \Sigma\, (-i)^j \dim H^j(D) .$$

In fact one would get the same result by replacing the spaces of holomorphic sections $O(X,E_j)$ by spaces of C^∞ or distribution sections. If D is almost elliptic we stick to the spaces of holomorphic sections ; such a section extends to X_ε for small ε . If D_ε is the extension of D to X_ε , it is elliptic for small $\varepsilon > 0$. so the $H^j(D_\varepsilon)$ are finite dimensional. It follows that they are constant and equal to their limit $H^j(D)$, so Ind D = Ind D_ε , for small $\varepsilon > 0$. In fact they remain constant as long as D_ε remains elliptic: [6] the restriction map $H^j(D_\varepsilon) \to H^j(D_{\varepsilon'})$ $(\varepsilon' < \varepsilon)$ is an isomorphism if D_α is elliptic for $\varepsilon' \leqslant \alpha \leqslant \varepsilon''$.

b. K-theory. Let X be a paracompact space, and Φ a family of supports on X . The K-theory with supports $K_\Phi(X)$ is defined [1] . Its elements are classes of complexes of vector bundles

$$(u) \qquad 0 \to E_o \xrightarrow{u} E_1 \xrightarrow{u} \dots E_n \to 0$$

where the differential u is exact outside of some set $A \in \Phi$. In fact the complex above is equivalent to the complex of length 2 :

$$0 \to \Sigma E_{2j} \xrightarrow{u+u^\star} \Sigma E_{2j+1} \to 0$$

where we choose some hermitian metrics on the E_j's to define the adjoints $u^\star$.

If $\Phi = \{Y\}$, whith Y a closed subset of X, we write $K_Y(X)$.

If Φ is the set of compact subsets of X, we write $K_c(X)$.

c. Virtual bundle defined by an elliptic system

If D is a complex of differential operators on a manifolds X, and $Z \subset T^\star X$ a closed set containing $\mathrm{char}(D)$, the symbol $\sigma(D)$ is exact outside of Z so it defines an element

(5) $$[D]_Z \in K_Z(T^\star X)$$

Suppose now that X is a compact complex manifold with boundary. Let ρ be a real function such that $\rho < 0$ on $X - \partial X$, $\rho = 0$ and $\partial\rho \neq 0$ on ∂X. Then $\partial\rho$ defines a section of $T^\star X$. If D is elliptic, this does not meet $\mathrm{char}(D)$ above ∂X so by restriction D defines an element of $K(X, X - \partial X) = K_c(X)$

(6) $$[D] \in K_c(X) \quad \text{if } D \text{ is elliptic, } X \text{ complex.}$$

Suppose now that X is real, compact, and D almost elliptic on X. Then D defines an element of $K_c(X_\varepsilon)$. For small $\varepsilon > 0$ the X_ε are all isomorphic to tubular neighbourhoods of X (they are tubular neighbourhoods of $\mathrm{Re}\, X_\varepsilon$). They are also isomorphic to the unit ball of $T^\star X$; in fact, up to homotopy, there is a unique isomorphism which is compatible with the almost complex structures : infinitesimally this will take the covector (X,ξ) into $X - i\varepsilon\xi$. Thus D defines a virtual bundle

(7) $$[D] \in K_c(T^\star X) \simeq K_c(X_\varepsilon) \quad \text{if } D \text{ is almost elliptic } X \text{ real, } \varepsilon \text{ small.}$$

d - The index character (topological index)

Let X be an even manifold. An almost complex (resp. symplectic) structure on X is a complex (resp. symplectic) structure on its (real) tangent bundle, without the integrability condition. Two such structures are homotopic if they can be linked by a continous one-parameter family of such structures.

There is a canonical one-to-one correspondance between homotopy classes of almost complex or almost symplectic structures : to an almost complex structure corresponds the non-degenerate 2-form $\omega(u,v) = \operatorname{Im} H(u,v)$, where H is any $\geqslant 0$ hermitian form (antilinear in the first variable).

Let X be a paracompact space and $N \overset{p}{\to} X$ a complex vector bundle on X . The pull back $p^\star N$ has a canonical section k $(n \to p^\star n)$. The Koszul complex of N is the complex K_N defined by the inner product by k on the exterior algebra $p^\star \Lambda N^\star$:

$$K_N : \ldots \to p^\star \Lambda^j N \xrightarrow{k_L} p^\star \Lambda^{j-1} N^\star \to \ldots p^{\star} N^\star \to \mathbb{C} \to 0$$

the graduation is such that $\deg (\Lambda^j N^\star) = - j$; $k_L(y) = k_L y$.

K_N is exact outside of the zero section $i : X \to N$, so it defines an element $[K_N] \in K_X(N)$. The Bott periodicity theorem can be restated as the fact that multiplication by $[K_N]$ $(\xi \to p^\star\xi . [K_N])$ induces an isomorphism $\beta_i : K_c(X) \to K_c(N)$.

Let X and Y be two real manifolds, $i : X \hookrightarrow Y$ and embedding. Suppose we are given a complex structure on the normal tangent bundle $TY/_{i(TX)}$. Since $i(X)$ has a tubular neighbourhood, we can transport the Bott map for the complex normal tangent bundle, and get a map

$$\beta_i : K_c(X) \to K_c(Y)$$

(this is usually no longer an isomorphism if Y itself is not a tubular neighbourhood ; it depends on the choice of the complex structure).

This can be applied in the two following cases :

i) X and Y are almost complex, and i is an almost complex embedding, then the normal complex bundle $TY/_{i(TX)}$ has the quotient complex structure.

ii) X and Y are almost complex, $p : Y \to X$ is an almost complex submersion and $i : X \hookrightarrow Y$ is a section of p . Then the normal tangent bundle is canonically isomorphic to the vertical tangent bundle $T_p Y/_{i(X)}$, which has the induced complex structure.

Let X be an almost complex (or almost symplectic) manifold. The topological index is a character

$$\chi_{top}^{X} : K_c(X) \to K_c(Y) .$$

It is characterised by the following conditions :

(χ1) if X = point, then χ_{top} : K(point) → **Z** is the dimension map.

(χ2) χ_{top}^{X} only depends on the homotopy class of the almost complex (or symplectic) structure of X .

(χ3) if $i : X \hookrightarrow Y$ is an almost complex embedding, then $\chi_{top}^{X} = \chi_{top}^{Y} \circ \beta_i$.

The topological index also has a cohomological expression $\chi_{top}^{X}(\xi) = \langle ch\ \xi , \tau_X \rangle$, where $Ch\ \xi$ is the Chern character of ξ , and τ_X the Todd class of the almost complex (or symplectic) structure of X .

We may now state the index theorem :

<u>Theorem</u> - <u>Let</u> D <u>be an almost elliptic system on a real or complex manifold</u> X <u>with boundary. Then</u>

$$\mathrm{Ind}(D) = \chi_{top}([D]) .$$

In the complex case X must be a Stein manifold, or the definition of the index must be modified ; we have $[D] \in K_c(X)$, and X has its given complex structure for the definition of χ_{top}.

In the real case we have $[D] \in K_c(T^{\star}X)$, and χ_{top} corresponds to the canonical symplectic structure of $T^{\star}X$. Equivalently $[D] \in K_c(X_\varepsilon)$ with notation as above ; the identification of $T^{\star}X$ and X_ε is made through a map tangent to $(X,\xi) \mapsto X - i\xi$ which is compatible with the complex and symplectic structures.

e - <u>Going from real to complex</u>

We have seen that the real case can be reduced to the complex one, replacing X by X_ε : this is compatible with the index, and with χ_{top} .

Conversely the complex case can be reduced to the real one, replacing X by X_R and D by $D_R = D \otimes \bar{\partial}$. This is compatible with the cohomology and the index if X is Stein. In fact if X is not Stein, this provides the right definition of the index : $\operatorname{Ind} D = \operatorname{Ind} D_R$: one should replace cohomology by hypercohomology, and $\bar\partial$ is a cohomologically trivial resolution of O_X .

Let us show that the topological index also follows in the process. The complex manifold $Y = X \times \bar X$ is a complexification of X_R ; the embedding $i : X_R \hookrightarrow Y$ is the diagonal map $x \mapsto (x,\bar x)$. Let us denote by (x,y) the variable in $Y = X \times \bar X$. We also choose a hermitian metric on X (and the same on $\bar X$). We may choose the functions u and ρ as in § 1 so that $u(x,y) = |y-\bar x|^2 + O(|y-\bar x|^3)$, and ρ is independent of y .

The holomorphic extension of D_R on Y is $D_x \otimes \partial_y$. Its symbol evaluated at $\partial(u+\varepsilon\rho)$ is $\sigma_D(\partial_x u + \varepsilon\partial_x\rho) \otimes g_y(\partial_y)$ it is clearly homotopic to

$$\sigma_D(\partial_x\rho) \otimes \sigma_{\partial y}(\partial_y u) .$$

Hence $[D_R] = p^\star[D] . \lambda \in K_c(X \times \bar X)$ (or $K_c(X_{R,\varepsilon})$ where p is the first projection $X \times \bar X \to X$, and $\lambda \in K_{X_R}(X \times \bar X)$ is the virtual bundle defined by the exterior product by $\partial_y u$ complex on $\Lambda T^\star \bar X$.

On the other hand $\overline{\partial_y u}$ can be viewed as a section of $T\bar X$ (identified with the antidual of $T^\star\bar X$ via the hermitian metric). Then it is clearly homotopic to the canonical section defined by a tubular neighbourhood of X_R in Y . Hence the Koszul complex of X_R in Y is homotopic to the inner product by $\partial_y \bar u$ complex on $\Lambda T^\star \bar X$.

Finally observe that the virtual bundles defined by $\partial_y u_\wedge$. and $\partial_y \bar u_\llcorner$. are the same : these two complexes are adjoint to each other, so the complexes of length two they define $(d+d^\star)$ are the same. So we finally see

$$[D_R] = p^\star[D] . \lambda = \beta_i [D] .$$

§ 3 - Embedding and proof of the index theorem

From now on it will be convenient to use the language of $\mathcal{D}$-modules, for which we refer to [7], [8]. Let X be a real or complex manifold, O_X the sheaf of analytic functions on X , $\mathcal{D}_X$ the sheaf of analytic differential operators : it is a coherent sheaf of filtered algebras. If e is an analytic vector bundle on X - i.e. a locally free (= projective) O_X-module, then $E = e \otimes_{Ox} \mathcal{D}_x$ is a locally free right $\mathcal{D}_x$-module ; the initial vector bundle e is canonically isomorphic with the quotient $E \otimes_{\mathscr{D}x} O_x$ of E ; thereby differential operators on e correspond exactly to $\mathscr{D}$-homomorphisms on E . Thus a complex d of differential operators on X corresponds to a complex D of locally free right $\mathscr{D}_x$-modules ($d = D \otimes_{\mathscr{D}x} O_x$) . If we are given an increasing filtration $e = \bigcup_{j \in Z} e_j$ of e , the corresponding good filtration of E is $E_j = \Sigma\, e_{j-k} \otimes \mathscr{D}_k$ (the E_j are coherent as O_x-modules, we have $E_{j} \mathscr{D}_k \subset E_{j+k}$ and both are equal for large j).

If M is a coherent right $\mathscr{D}$-module with a good filtration on X [8] , the symbol $\sigma(m) = \mathrm{Gr}\, M$ is a coherent homogeneous $O_{T^\star X}$ - modulo on $T^\star X$. The singular spectrum $SS\, M \subset T^\star X$ is the support of $\sigma(M)$; it is a closed involutive analytic cone. The support supp M is the projection of SS M on X.

If $\dot{M}$ is a complex of coherent right $\mathscr{D}$-modules, the homology sheaves $\mathscr{H}^j(\dot{M})$ are coherent right $\mathscr{D}$-modules. If the M_j posess good filtrations, these can be modified so that for the induced filtrations one has

$$\sigma(\mathscr{H}^j(\dot{M})) = \mathscr{H}^j(\sigma(\dot{M}))$$

(good complex filtration). We denote by $SS\,\dot{M}$ the singular support of $\mathscr{H}^\star(\dot{M})$: it is again a closed analytic involutive cone in $T^\star X$.

The symbol $\sigma(\dot{M})$ depends on the choice of good filtrations. However if $Z \supset SS\,\dot{M}$, the element $[\dot{M}] \in K_Z(T^\star X)$ that it defines depends only on $\dot{M}$. If $\dot{M}$ corresponds to a complex D of differential operators, we always have $SS\,\dot{M} \subset \mathrm{char}(D)$, and $[\dot{M}] = [D] \in K_Z(T^\star X)$ if $Z \supset \mathrm{char}(D)$; but one may have $SS\, M \neq \mathrm{char}\,(D)$ if the canonical filtration corresponding to D is not a good complex filtration, and in this case char (D) is possibly not involutive -this has no consequence for index computations.

We may extend to $\mathscr{D}$-modules the definitions of ellipticity or almost ellipticity of § 1 , replacing char (D) by SS(M).

If $\dot{M}$ is a complex of coherent right $\mathscr{D}$-modules with good filtrations on X , the homology groups $H^j(X,\dot{M})$ are defined as the hypertorsion groups of $\dot{M} \otimes_{\mathscr{D}x} O_x$. If X is a Stein manifold , eg. if it is real, then $\dot{M}$ is quasi-isomorphic to a locally free complex $\dot{N}$ corresponding to a complex D of differential operators ($D = \dot{N} \otimes_{\mathscr{D}x} O_x$) ; then $H^\star(X,\dot{M}) = H^\star(X,\dot{N})$ is the homology of D . If X is not Stein one must also put in a homologically trivial resolution of O_X : (a canonical one is the Cauchy-Riemann complex $\bar{\partial}$), so we replace X by X_R , $\dot{M}$ by $\dot{M}_R = \dot{M} \otimes \bar{\partial}$ and set $H^j(X,\dot{M}) = H^j(X_R, \dot{M}_R)$.

If $\dot{M}$ is almost elliptic and has compact support, in particular if X is compact, then its homology groups are finite. Here again, its symbol defines an element $[\dot{M}] \in K_c(X)$ if X is complex, or $K_c(T^{\star}X)$ if X is real.(†) The index theorem may be restated as

Theorem bis - Let $\dot{M}$ be a compactly supported, almost elliptic complex of coherent right $\mathscr{D}$-modules with good filtrations. Then

$$\text{Ind.}(\dot{M}) = \Sigma(-1)^j \dim H^j(\dot{M}) = \chi_{top}([\dot{M}]).$$

To prove the theorem we need one last ingredient. Let Y be an analytic manifold, X an analytic submanifold, i the inclusion $X \hookrightarrow Y$. O_X may be viewed as a coherent O_Y-module ; the virtual bundle $[O_X] \in K_X(Y)$ it defines is precisely the element defined by the Koszul complex.

Let us denote by $\mathscr{D}_{XY}$ the sheaf of differential operators from Y to X , i.e. of the form $f \mapsto Pf|x$, where P is some differential operator. It is a $\mathscr{D}_X$, $\mathscr{D}_Y$ bi-module, and coherent on $\mathscr{D}_Y$, and one has obviously

$O_X \simeq \mathscr{D}_{XY} \otimes_{\mathscr{D}_Y} O_Y$, $\mathscr{D}_{XY} \simeq O_X$.

If M is a right $\mathscr{D}_X$-module and $i_{\star}M$ its sheaf theoretic image on Y , the direct image $i_!M$ is the right $\mathscr{D}_y$-module

$$i_!M = i_{\star}M \otimes_{\mathscr{D}x} \mathscr{D}_{XY}.$$

For example, let $X = \mathbb{R}^p$ (or $\mathbb{C}^p$) , $Y = \mathbb{R}^{p+q}$ (resp. $\mathbb{C}^{p+q}$).

Let $k : 0 \to \Lambda^q\, \mathbb{C}^q \to \Lambda^{q-1}\, \mathbb{C}^q \to \dots\, \mathbb{C}^q \to \mathbb{C} \to 0$ be the Koszul complex of $\{0\}$ in $\mathbb{C}^q$, and $\delta = k \otimes_{O_{\mathbb{C}^q}} \mathscr{D}_{\mathbb{C}^q}$: this may be viewed as a projective resolution of the $\mathscr{D}$-module of distributions carried by the origin (Dirac mass

---

(†) $[\dot{M}]$ depends only on $\dot{M}$, and not on the choice of good filtrations, resolutions, etc... .

and its derivative); it is holonomic hence almost elliptic. If M is a right $\mathscr{D}_X$-module, $i_! M$ is isomorphic to the external product

$$i_! M = M \otimes \delta .$$

It follows that $i \dot{M}$ is almost elliptic if $\dot{M}$ is : this is true if $X = \mathbb{R}^p \subset \mathbb{R}^{p+q}$ by prop. 2, and in general because almost ellipticity is a local property anyway.

It follows also from the fact that the symbol of a resolution of $\mathscr{D}_{XY}$ is the Koszul complex that we have

$$[i_! M] = \beta_i [M] \quad \text{if } M \text{ is almost elliptic.}$$

One checks also trivially that the homology sheaves $\mathcal{H}^j(i_! \dot{M})$ and $\mathcal{H}^j(\dot{M})$ are the same, so

$$H^j(Y, i_! \dot{M}) = H^j(X, \dot{M}) \quad \text{and} \quad \mathrm{Ind}(i_! \dot{M}) = \mathrm{Ind}\,(\dot{M}) .$$

We may now prove the index theorem. We first need to know that $\mathrm{Ind}\,\dot{M}$ only depends on $[\dot{M}]$. Since $\mathrm{supp}\,\dot{M}$ is compact we may suppose that X is compact with boundary, replacing it by a suitable compact neighborhood of $\mathrm{supp}\,\dot{M}$ if need be. If X is complex but not Stein, we may replace X by X_R and $\dot{M}$ by $\dot{M}_R$. If X is real we may replace it by X_ε for some small ε : in all these replacements, index and χ_{top} are unchanged. We are thus reduced to the case where X is a complex compact strictly pseudo-convex manifold with boundary. In this case we dispose of the theory of Toeplitz operators on ∂X [3] , for which we have symbolic calculus, adjoints etc..., as for pseudodifferential operators. If $\dot{M}$ corresponds to a complex $D : \ldots \to E_j \to E_{j+1} \to \ldots$ of differential operators, we may define the Toeplitz operator

$$D + D^\star \; : \; \Sigma\, E^{2j} \to \Sigma\, E^{2j+1} .$$

This is elliptic if D is elliptic ; we have $[D + D^\star] = [D]$ in $K_c(X) = K(X, \partial X)$, and $\mathrm{Ind}\,(D + D^\star) = \mathrm{Ind}\,(D)$ by Hodge theory. Now clearly $\mathrm{Ind}\,(D + D^\star)$ as $\mathrm{Ind}\,(D)$ is additive, and is invariant under small perturbations as the index of any Fredholm operator, so it depends only on the homotopy class of the symbol, i.e. on $[D]$.

If X is a point, a complex of $\mathcal{D}_X$-modules is just a finite complex D : $\ldots \to E_j \to E_{j+1} \to \ldots$ of finite dimensional vector spaces, so in this case we have

$$\mathrm{Ind}(D) = \Sigma(-1)^j \dim E_j = \chi_{top}[D] .$$

If $X = \mathbb{R}^n$ or the unit ball $B_n \subset \mathbb{R}_n$, so $T^\star X = \mathbb{R}^{2n} = \mathbb{C}$ (resp. $T^\star X = X \times \mathbb{R}^n \subset \mathbb{C}^n$), and $\dot{M}$ is an almost elliptic complex of coherent right $\mathcal{D}$- modules with good filtrations, we have seen that $\mathrm{Ind}\,\dot{M}$ depends only on $\dot{M}$. If i is the obvious inclusion $\{0\} \to \mathbb{R}^n$ and $\dot{M} = i_!\dot{N}$, then $\mathrm{Ind}\,\dot{M} = \mathrm{Ind}\,\dot{N} = \chi_{top}[\dot{N}] = \chi_{top}\,\beta_i[\dot{N}] = \chi_{top}[\dot{M}]$. Since $\beta_i : K(\text{point}) \to K_c(T^\star X)$ is an isomorphism by Bott's theorem, the index theorem is proved in this case.

If X is real and compact, we may find an embedding $i : X \hookrightarrow B_n$ such that $i(\partial X) \subset \partial B_n$ and $i^{-1}(\partial B_n) = \partial X$. Here again we have

$\mathrm{Ind}\,\dot{M} = \mathrm{Ind}\, i_!\dot{M} = \chi_{top}[i_!\dot{M}] = \chi_{top}[\dot{M}]$.

Finally if X is complex we have

$\mathrm{Ind}\,\dot{M} = \mathrm{Ind}\,\dot{M}_R = \chi_{to}[\dot{M}_R] = \chi_{top}[\dot{M}]$.

This proves the index theorem in all cases. As a particular case one gets back the Riemann-Roch theorem for a complex of coherent sheaves ξ on a compact analytic space Z embeddable in a complex manifold X ($i : z \hookrightarrow X$) :

$$\chi(F) = \chi(i_\star\xi) = \mathrm{Ind}\,(i_\star\xi \otimes_{O_x} \mathcal{D}_x) = \chi_{top}[i_\star\xi] = \chi_{top}[\xi]$$

(the corresponding $\mathcal{D}_X$-module is elliptic if ∂X does not meet supp ξ).

BIBLIOGRAPHY

[1] M.F. ATIYAH : K-theory, Benjamin (Amsterdam) 1967.

[2] M.F. ATIYAH, C. SINGER : The index of elliptic operators. Ann. Math. 87 (1968), 489-530.

[3] L. BOUTET de MONVEL : The index of Toeplitz operators. Inventiones Math. 50 (1979) 249-272.

[4] L. BOUTET de MONVEL - B. MALGRANGE : Systèmes presqu'elliptiques... Colloque L. Schwartz, Orsay 1983.

[5] A. GROTHENDIECK : S.G.A. 5

[6] M. KASHIWARA : Cours sur les équations microdifferentielles.

[7] M. KASHIWARA - T. KAWAI - M. SATO : Microfunctions and pseudo-differential equations. Lectue Notes 287, chap. II, Springer 1973.

[8] B. MALGRANGE : Séminaire sur les opérateurs différentiels et pseudo-différentiels. Grenoble 1975-76.

L. BOUTET de MONVEL
Département de Mathématiques
Université PARIS VI
4, Place Jussieu
75005 - PARIS

L CATTABRIGA

Continuation of regularity and surjectivity of differential operators in Gevrey-spaces

In this talk we report on certain results that developed from the ideas contained in the works [6], [7], [8] of E De GIORGI and the author. We are concerned with problems relating to linear partial differential operators with constant (except in Section 3) coefficients acting on the so-called Gevrey spaces, which we now describe briefly.

1 - Gevrey Spaces (1)

$\gamma^{(\rho)}(K)$, $\rho > 1$, K a compact set of R^n, will denote the linear space of all complex valued functions $\varphi \in C^\infty(R^n)$ such that for every $\varepsilon > 0$

$$(1.1)\quad \|\varphi\|_{\varepsilon,K} = \sup_{\alpha\in\mathbb{Z}^n_+} \sup_{x\in K} \Gamma(\rho|\alpha|+1)^{-1}\, \varepsilon^{-|\alpha|}\, |D^\alpha \varphi(x)| < \infty .$$

Here Γ is the Euler gamma function, $D = (D_1,\dots,D_n)$, $D_j = -i\partial/\partial x_j$.

$|\alpha| = \alpha_1+\dots+\alpha_n$. As is well known $\gamma^{(\rho)}(R^n) = \bigcup_K \gamma^{(\rho)}(K)$ is a Fréchet-Montel space with family of seminorms (1.1) ; moreover, starting from the space $\gamma_0^{(\rho)}(R^n) = \gamma^{(\rho)}(R^n) \cap C_0^\infty(R^n)$ endowed with natural topology of inductive limit of the spaces $\gamma^{(\rho)}(K)$, one can define the space of Beurling ultradistributions $\gamma_0^{(\rho)'}(R^n)$, as the dual space of $\gamma_0^{(\rho)}(R^n)$. The space $\gamma^{(\rho)'}(R^n)$ defined as the dual space of $\gamma^{(\rho)}(R^n)$ can be identified with the space of the ultradistributions in $\gamma_0^{(\rho)'}(R^n)$ with compact support.

We shall also consider the space $\Gamma^{(\sigma)}(\Omega)$, $\sigma \geq 1$, Ω an open set of R^n, of all complex valued functions $f \in C^\infty(\Omega)$ such that for every compact set $K \subset \Omega$ there exists a constant c such that :

---

(1) See for example [11] .

$\xi \in R^n, t \in R$, $|\xi - <\xi, N^j> N^j| > k$, $P(\xi + i t N^j) = 0 \Longrightarrow$

either $t \geqslant - c_1 |\xi - <\xi, N^j> N^j|^{1/\rho}$ or $t \leqslant - c_2 |\xi|^{1/\sigma}$;

ii) for every $y \in V \setminus \{0\}$ there exists $j \in \{1,\dots,h\}$ such that $<y, N^j> \gg 0$.

Then there exists a finite number of cones $\Delta^j = \{x \in R^n;$ $|x| \leqslant d_j < x, N^j > \}$, $d_j > |N^j|^{-1}$, $j = 1,\dots,h$ such that $V \subset \bigcup_{j=1}^{h} \Delta^j$ and for every $j = 1,\dots,h$

$u \in \gamma_o^{(\rho)'}(R^n)$, $u \in \Gamma^{(\sigma)}(C\Delta^j)$, $P(D) u \in \Gamma^{(\sigma)}(R^n) \Longrightarrow u \in \Gamma^{(\sigma)}(R^n)$.

3 - A result for operators with variable coefficients

A parametrix of the same type as the fundamental solution considered in Theorem 2.1 can also be obtained in some cases when P has variable coefficients, by means of a pseudo-differential operator of infinite order acting on Gevrey spaces. This new class of pseudo-differential operators has been defined and studied by L. ZANGHIRATI [14] . As an example consider the operator

$$(3.1) \qquad D_n^m + \sum_{j=o}^{m-1} a_j(x', D') D_n^j ,$$

where $a_j(x', D')$ are pseudo-differential operators of order $m_j \leqslant p(m-j)$, $p \in]0,1[$, acting on $\Gamma_o^{(\sigma)}(\Omega')$, Ω' an open subset of R^{n-1} , $\sigma \in [1, 1/p[$. In this case the following result can be proved with the aid of a parametrix of the above mentioned type.

Theorem 3.1(5)

Let P be the operator given by (3.1) and let

$a_j \in \Gamma^{(\sigma)'}(\Omega')$, $j = 0,\dots,m-1$,

$f(x', x_n) \in C^o(R^+; \Gamma^{(\sigma)'}(\Omega'))$. Then there exists a solution

$u \in C^{m-1}(R; \Gamma^{(\sigma)'}(\Omega'))$ of the Cauchy problem

(5) For the case when the a_j's do not depend on x', see [14]

$$\begin{cases} P(x',D)u = f & \text{in} \quad \Omega' \times R^+ \\ D_n^j u(x',o) = g_j(x') & x' \in \Omega', \; j = 0,\dots,m-1, \end{cases}$$

and for every $x_n > 0$

$$\sigma\text{-sing supp } u(x',x_n) \subset \left(\overline{\bigcup_{o \le s \le x_n} \sigma\text{-sing supp } f(x',s)} \right) \cup \left(\bigcup_{j=o}^{m-1} \sigma\text{-sing supp } g_j \right),$$

where σ-sing supp $u(x', x_n)$ denotes the subset of points in Ω' having no neighborhood in which the restriction of $u(.,x_n)$ is in $\Gamma^{(\sigma)}$.

4 - We want now to show how condition (2.2) is also sufficient in order for the equation $P(D)u = f$ to have a solution in $\Gamma^{(\sigma)}(R^n)$ for certain $f \in \Gamma^{(\sigma)}(R^n)$. For more details see [1] and [4].

We make use of the representation formula for functions in $\Gamma^{(\sigma)}(R^n)$, σ a rational number $\geqslant 1$, given by the following theorem.

Theorem 4.1 [1] (6)

Let $f \in \Gamma^{(\sigma)}(R^n)$, $\sigma = p/q \geqslant 1$, p,q positive integers and let χ be a positive non-increasing function on R. Then there exists a function $g \in C^\infty(R^n \times R^+)$ and a positive non-increasing function $\psi \in C^\infty(R)$ such that

$$(4.1) \qquad f(x) = \int_{R^n} dy \int_o^{+\infty} G(x-y,r)\, g(y,r)\, dr \qquad x \in R^n,$$

$$\text{supp } g \subset \{(y,r) \in R^n \times R^+ \; ; \; \psi(|y|^2) \leqslant r \leqslant 2\, \psi(|y|^2)\} ,$$

$$\int_o^{+\infty} |g(y,r)|\, dy \leqslant \chi(|y|^2) ,$$

where $G(x,r) = E(x,t)\Big|_{|t|=r}$ and

$$E(x,t) = \int_o^{+\infty} \mathcal{F}^{-1}_{(\xi,\tau)}(\exp(-\nu Q(\xi,\tau)))\,(x,t)\, d\nu ,\text{(7)}$$

$t \in R^s$, is a fundamental solution on R^{n+s} of the operator

$$Q(D_x,D_t) = \left(\sum_{j=1}^{n} D_{x_j}^2\right)^q + \left(\sum_{h=1}^{s} D_{t_h}^2\right)^p , \quad s > 2\,p.$$

(6) When $\sigma = 1$ see [7].

(7) Here $\mathcal{F}^{-1}_{(\xi,\tau)}$ denotes the inverse Fourier transformation with respect to $(\xi,\tau) \in R^{n+s}$.

Thus, letting $E(x,t;\nu) = \mathcal{F}^{-1}_{(\xi,\tau)}(\exp(-\nu Q(\xi,\tau)(x,t)$ and for $\nu_o > 0$

$$G_1(x,r) = \int_o^{\nu_o} E(x,t;\nu)\Big|_{|t|=r} d\nu\ , \quad G_2(x,r) = \int_{\nu_o}^{+\infty} E(x,t;\nu)\Big|_{|t|=r} d\nu\ ,$$

a function $f \in \Gamma^{(\sigma)}(R^n)$ can be written as a sum $f_1 + f_2$

where

$$f_i(x) = \int_{R^n} dy \int_o^{+\infty} G_i(x-y,r)\ g(y,r)\ dr\ ,\ i = 1,2,\ x \in R^n\ .$$

Let now P be a differential operator satisfying condition (2.2). If we want to find a solution $u \in \Gamma^{(\sigma)}(R^n)$ of the equation $P(D)u = f$, for a given $f \in \Gamma^{(\sigma)}(R^n)$, we first note that the function f_2 defined above is in $\gamma^{(\sigma)}(R^n)$, for every $\nu_o > 0$. Since $\gamma^{(\sigma)}(R^n) \subset \Gamma^{(\sigma)}(R^n)$ and $P(D)\ \gamma^{(\sigma)}(R^n) = \gamma^{(\sigma)}(R^n)$ for every differential operator P with constant coefficients and for every $\sigma \geqslant 1$ [(8)] , we are left to show the existence of a solution $u_1 \in \Gamma^{(\sigma)}(R^n)$ of the equation $P(D)u_1 = f_1$ for a suitable ν_o.

To this end we first find a function $H(.,r) \in C^\infty(R^n)$ such that for a suitable ν_o

$$P(D)\ H(x,r) = G_1(x,r)\ , \quad x \in R^n\ ,\ r > 0$$

and for every $\alpha \in \mathbf{Z}^n_+$

(4.2) $\quad |D^\alpha_x\ H(x.r)| \leqslant c^{|\alpha|+1}\ e^{c'|x|}\ r^{-\sigma|\alpha|}\quad \Gamma(\sigma|\alpha|+1)\ F(|x_n|,r)$

for every $x \in R^n$ and

(4.2') $\quad |D^\alpha_x\ H(x,r)| \leqslant c^{|\alpha|+1}\ e^{c'|x|}\quad \Gamma(\sigma|\alpha|+1)\quad$ when $\quad x_n \leqslant -s/2p$,

where c,c' are positive constants and F is a continuous positive function such that $\lim\limits_{r \to o} F(|x_n|,r) = +\infty$.

Thus putting

$$u_1(x) = \int_{R^n} dy \int_o^{+\infty} H(x-y,r)\, g(y;r)\ dr\quad ,\quad x \in R^n$$

we prove the following theorem :

---

(8) See for example [12] .

Theorem 4.2 [1],[5] when $\sigma = 1$.

Let σ be a rational number $\geqslant 1$ and let P, given by (2.1), satisfy condition (2.2). Suppose that $f \in \Gamma^{(\sigma)}(R^n)$ be such that for a certain choice of the function χ in Theorem 4.1, the function g in (4.1) satisfies the conditions

(4.3) $\quad \int_R \exp(c'|y|)\, dy \int_0^{+\infty} |g(y,r)|\, dr < \infty ,$

$\operatorname{supp} g \subset \{(y,r);\ |y| \leqslant d\, y_n ,\quad \psi(|y|^2) \leqslant r \leqslant 2\,\psi(|y|^2)\} ,$

where c' is the same constant as in (4.2),(4.2') and $d > 1$. Then there exists a solution $u \in \Gamma^{(\sigma)}(R^n)$ of the equation $P(D)\,u = f$.

From this theorem the following corollary immediately follows.

Corollary 4.3

The conclusion of Theorem 4.2 holds when P is a differential operator (with constant coefficients) satisfying condition (2.3) of Corollary 2.4 for a certain closed cone $V \subset R^n$ and the function $f \in \Gamma^{(\sigma)}(R^n)$ is such that the function g in (4.1) satisfies (4.3) and

(4.4) $\quad \operatorname{supp} g \subset \{(y,r);\ y \in V ,\ \psi(|y|^2) \leqslant r \leqslant 2\,\psi(|y|^2)\} .$

Note that condition (4.3) can always be satisfied by choosing the function χ in Theorem 4.1 such that

$$\int_R \exp(c'|y|)\,\chi(|y|^2)\,dy < \infty .$$

According to Theorem 4.1, condition (4.4) is empty when $V = R^n$. As a consequence, Corollary 4.3 gives

Theorem 4.4 [1]

Let P be a linear differential operator with constant coefficients and suppose that P satisfies condition (2.3) with $V = R^n$ and σ a rational number $\geqslant 1$. Then $P(D)\,\Gamma^{(\sigma)}(R^n) = \Gamma^{(\sigma)}(R^n)$.

When $\sigma = 1$ and $\rho = m/(m-1)$ the conditions required by this theorem are in particular satisfied by every P of order m with hyperbolic-elliptic principal part. The same conditions are satisfied for every rational number $\sigma \geqslant 1$, by every differential operator P in two variables (9). On the other hand, it has been proved in [2] that there exist differential operators P on R^n, $n \geqslant 3$, such that $P(D)\,\Gamma^{(\sigma)}(R^n) \neq \Gamma^{(\sigma)}(R^n)$, for some $\sigma \geqslant 1$. For $n = 3$ this is the case for example when $P(D) = -D_1^2 - D_2^2 - i\,D_3$ and $\sigma \in [1,2[$ and when $P(D) = -D_1^2 - D_2^2$ or $P(D) = i\,D_1 - D_2$ and any $\sigma \geqslant 1$ (10).

---

(9) For $\sigma = 1$ see [8]. The equality $P(D)\,\Gamma^{(\sigma)}(R^2) = \Gamma^{(\sigma)}(R^2)$ for every operator P in two variables and every real $\sigma \geqslant 1$ has been subsequently proved in [13].

(10) For $\sigma = 1$ see [6] and the necessary and sufficient conditions proved in [9].

REFERENCES

[1] L. CATTABRIGA : Solutions in Gevrey spaces of partial differential equations with constant coefficients, Astérisque, Soc. Math. de France 89-90 (1981), 129-151.

[2] L. CATTABRIGA : On the surjectivity of differential polynomials on Gevrey-spaces, Rend. Sem. Mat. Univ. Politecn. Torino, special issue 1983, 81-89.

[3] L. CATTABRIGA : Some remarks on continuation of Gevrey regularity for solutions of linear partial differential equations with constant coefficients, Rend. Sem. Mat. Fis. Milano (to appear).

[4] L. CATTABRIGA : Alcuni problemi per equazioni differenziali lineari con coefficienti costanti, Quaderni dell'Unione Matematica Italiana, n° 24, 1983.

[5] L. CATTABRIGA - E. DE GIORGI, Soluzioni di equazioni differenziali a coefficienti costanti appartenenti in un semispazio a certe classi di Gevrey, Boll. Un. Mat. Ital. (4) 12 (1975), 324-348.

[6] E. DE GIORGI : Solutions analytiques des équations aux dérivées partielles à coefficients constants, Séminaire GOULAOUIC-SCHWARTZ, 1971 - 1972, exposé n° 29.

[7] E. DE GIORGI - L. CATTABRIGA : Una formula di rappresentazione per funzioni analitiche in R^n , Boll. Un. Mat. Ital. (4) 4 (1971), 1010 - 1014.

[8] E. DE GIORGI - L. CATTABRIGA : Una dimostrazione diretta dell'esistenza di soluzioni analitiche nel piano reale di equazioni a derivate parziali a coefficienti costanti, Boll. Un. Mat. Ital. (4) 4 (1971), 1015 - 1027.

[9] L. HÖRMANDER : On the existence of real analytic solutions of partial differential equations with constant coefficients, Inventiones Math. 21 (1973), 151-182.

[10] L. HÖRMANDER : The Analysis of Linear Partial Differential Operators II, Springer-Verlag, 1983.

[11] H. KOMATSU : Ultradistributions I . Structure theorems and a characterization, J. Fac. Sci. Univ. Tokyo, Sec. I A Math. 20 (1973), 25-105.

[12] F. TREVES : Locally Convex Spaces and Linear Partial Differential Equations, Springer Verlag, 1967.

[13] G. ZAMPIERI : Risolubilità negli spazi di Gevrey di operatori differenziali di tipo iperbolico-ipoellittico (to appear).

[14] L. ZANGHIRATI : Pseudo-differential operators of infinite order and Gevrey classes (to appear).

L. CATTABRIGA
Dipartimento di Matematica
Universita di Bologna
Piazza di Porta S Donato 5
40127 Bologna (Italia)

G DAL MASO

Some singular perturbation problems in the calculus of variations

I shall consider some singular perturbation problems in the calculus of variations which arise from optimal control theory.

Let Ω be a bounded open subset of $\mathbb{R}^n$ with a smooth boundary $\partial\Omega$ and let $p \in \mathbb{R}$ with $2 \leqslant p < +\infty$. The optimal control problem I am going to consider is the following : minimize the cost functional

$$J(u,v) = N \int_\Omega |v|^2 dx + \int_\Omega |u-b|^p dx$$

among all pairs $(u,v) \in L^p(\Omega) \times L^2(\Omega)$ which satisfy the state equation

$$(E_\varepsilon) \qquad \begin{cases} \varepsilon \Delta u + g(u) = v \\ \varepsilon u \in H^1_o(\Omega) . \end{cases}$$

Here N is a constant, $N > 0$, b is a function in $L^p(\Omega)$, ε is a parameter, $\varepsilon \geqslant 0$, and $g : \mathbb{R} \to \mathbb{R}$ is a continuous function with the following properties :

(1) $\qquad |g(s)| \leqslant c(1+|s|^{\frac{p}{2}}) \qquad \forall s \in \mathbb{R}$

(2) $\qquad |g(t) - g(s)| \leqslant \rho(|t-s|)(1+|s|^{\frac{p}{2}}) \qquad \forall s \in \mathbb{R} \quad \forall t \in \mathbb{R}$

where c is a constant and $\rho : [0,+\infty[\to [0,+\infty[$ is increasing, continuous, with $\rho(0) = 0$.

Let us define for every $\varepsilon \geqslant 0$

$$J_\varepsilon = \inf \{J(u,v) : (u,v) \in L^p \times L^2, (u,v) \text{ satisfies } (E_\varepsilon) \} .$$

I shall investigate the asymptotic behaviour of J_ε as $\varepsilon \to 0_+$. Two questions are natural :

(A) Does there always exist $\lim\limits_{\varepsilon \to 0_+} J_\varepsilon$?

(B) Under which conditions $\lim\limits_{\varepsilon \to 0_+} J_\varepsilon = J_o$?

This problem has been proposed by J.L. LIONS and has been studied by A. BENSOUSSAN [1], A. HARAUX and F. MURAT [6], [7], [8], and V. KORMORNIK [9].

I shall illustrate here a different approach to this problem, which has been developed by G. BUTTAZZO and myself (see [2], [3]), and which relies on De Giorgi's Γ-convergence.

The $\Gamma(X^-)$-convergence is a convergence for functions defined on a topological space X and with values on $\overline{\mathbb{R}}$. One of the most remarkable properties of $\Gamma(X^-)$-convergence is given by the following theorem (see [4]).

Theorem 1

If the sequence (f_h) $\Gamma(X^-)$-convergences to f, and if the set

$$\bigcup_{h=1}^{\infty} \{x \in X : f_h(x) \leqslant t\}$$

is conditionally compact in X for every $t \in \mathbb{R}$, then

$$\inf_{x\in X} f(x) = \lim_{h\to\infty} \inf_{x\in X} f_h(x) .$$

In order to apply Γ-convergence, it is convenient to transform our optimal control problem into an ordinary variational problem. This is very easy in this case, because from the state equation (E_ε) we can write v in terms of u; thus, if we substitute $v = \varepsilon \Delta u + g(u)$ in the cost functionnal, we obtain :

$$J_\varepsilon = \inf_{u\in U} \int_\Omega [N|\varepsilon \Delta u + g(u)|^2 + |u-b|^p]\, dx$$

where $U = \{u \in L^p(\Omega) : \varepsilon u \in H^2(\Omega) \cap H^1_o(\Omega)\}$. Let us define $f : \Omega \times \mathbb{R} \times \mathbb{R} \to \mathbb{R}$ by

$$f(x,s,z) = N|z + g(s)|^2 + |s-b(x)|^p$$

and let $F_\varepsilon : L^p(\Omega) \to [0,+\infty]$ be defined by

$$F_\varepsilon(u) = \begin{cases} \int_\Omega f(x,u(x),\varepsilon \Delta u(x))dx & \text{if } \varepsilon u \in H^2(\Omega) \cap H^1_o(\Omega) , \\ +\infty & \text{otherwise.} \end{cases}$$

Then for every $\varepsilon \geqslant 0$

$$J_\varepsilon = \inf_{u \in L^p(\Omega)} F_\varepsilon(u).$$

Let us denote by $w\text{-}L^p(\Omega)$ the space $L^p(\Omega)$ endowed with the weak topology. Then the following theorem holds.

Theorem 2

There exists a function $\psi(x,s)$, convex in s, such that the sequence $(F_\varepsilon)_{\varepsilon>o}$ $\Gamma(w\text{-}L^p(\Omega)^-)$-converges (as $\varepsilon \to 0_+$) to the functional

$$\Psi(u) = \int_\Omega \psi(x,u(x))dx .$$

By theorem 1 this implies that

$$\inf_{u\in L^p(\Omega)} [\Psi(u) + \int_\Omega \varphi\, u dx] = \lim_{\varepsilon\to o_+} \inf_{u\in L^p(\Omega)} [F_\varepsilon(u) + \int_\Omega \varphi\, u dx]$$

for every $\varphi \in L^q(\Omega)$ $(p^{-1} + q^{-1} = 1)$. In particular for $\varphi = 0$ we obtain

$$\inf_{u\in L^p(\Omega)} \Psi(u) = \lim_{\varepsilon\to o_+} J_\varepsilon .$$

Therefore question (A) has an affirmative answer.

If we were able to prove that

$$\psi(x,s) = f(x,s,o)$$

then we would obtain

$$J_o = \inf_{u\in L^p(\Omega)} \Psi(u) = \lim_{\varepsilon\to o_+} J_\varepsilon .$$

So the answer to question (B) depends on the equality $\psi(x,s) = f(x,s,0)$.

The following theorem gives three formulae that can be used to compute ψ.

Theorem 3

For a.a. $x \in \Omega$ and for all $s \in \mathbb{R}$ we have

(a) $\psi(x,s) = \lim_{\varepsilon\to o_+} \inf \{F_\varepsilon(x,u) : u \in H^2(Y)\ ,\ \int_Y u dy = s\}$

(b) $\psi(x,s) = \inf \{F_\varepsilon(x,u) :\ \varepsilon > 0,\ u-s \in H^2_o(Y),\ \int_Y u dy = s\}$

(c) $\psi(x,s) = \inf \{ F_\varepsilon(x,u) : \varepsilon > 0,\ u \in H^2_\#(Y) ,\ \int_Y u\,dy = s \}$

where Y denotes the unit cube $]0,1[^n$, $H^2_\#(Y)$ denotes the space of all Y-periodic functions of $H^2_{loc}(\mathbb{R}^n)$, and

$$F_\varepsilon(x,u) = \int_Y f(x,u(y),\varepsilon\,\Delta u(y))\,dy .$$

The calculation of ψ by means of these formulae is not very simple, and it may seem as difficult as the original question (B). Nevertheless, we can get some useful information from theorem 3 , which allows us to determine explicitly ψ in some particular cases.

For instance, using theorem 3 we can obtain two fundamental inequalities concerning ψ . In order to state these inequalities, we need two definitions. Let us denote by

$$(co_s f)(x,s,z)$$

the convexification of f with respect to s , that is the greatest function convex in s which is less than or equal to f .

Let us denote by

$$(co_{s,z} f)(x,s,z)$$

the convexification of f with respect to (s,z), that is the greatest function convex in (s,z) which is less than or equal to f .

It is easy to see that

$$co_{s,z} f \leqslant co_s f \leqslant f .$$

Using formula (b) of theorem 3 we obtain the following inequalities.

<u>Proposition 4</u>

For a.a. $x \in \Omega$ and for all $s \in \mathbb{R}$

$$(co_{s,z} f)(x,s,0) \leqslant \psi(x,s) \leqslant (co_s f)(x,s,0).$$

This proposition has an immediate corollary.

<u>Corollary 5</u>

It for a.a. $x \in \Omega$ and for all $s \in \mathbb{R}$

$$(co_{s,z} f)(x,s,0) = (co_s f)(x,s,0) ,$$

then

$$\psi(x,s) = (co_s f)(x,s,0)$$

for a.a. $x \in \Omega$ and for all $s \in \mathbb{R}$.

Since

$$\inf_{u\in L^p(\Omega)} \int_\Omega (co_s f)(x,u(x),0)\,dx = \inf_{u\in L^p(\Omega)} \int_\Omega f(x,u(x),0)\,dx ,$$

(see [5], corollary 1.3, page 270), in the situation described by the corollary we have

$$J_0 = \inf_{u\in L^p(\Omega)} \int_\Omega f(x,u(x),0)\,dx = \inf_{u\in L^p(\Omega)} \Psi(u) = \lim_{\varepsilon\to 0_+} J_\varepsilon ,$$

and this is an answer to question (B).

This corollary can be applied in many different cases, that we shall consider in the following examples.

Example 6

Suppose that $f(x,s,z)$ is convex, in (s,z). Then $co_{s,z} f = co_s f = f$, thus

$$\psi(x,s) = f(x,s,0)$$

and $\lim_{\varepsilon\to 0_+} J_\varepsilon = J_0$. This happens, for instance, when $g(s)$ is an affine function.

Example 7 (See also [9]).

Suppose that g is convex and non-negative (or concave and non-positive). Then the function $f(x,s,z)$ is, in general, not convex in (s,z), but it is not difficult to show that

$$(co_{s,z} f)(x,s,0) = (co_s f)(x,s,0) = f(x,s,0)$$

for every $x \in \Omega$, $s \in \mathbb{R}$, thus

$$\psi(x,s) = f(x,s,0)$$

and $\lim_{\varepsilon\to 0_+} J_\varepsilon = J_0$. The function $g(s) = |s|^{\frac{p}{2}}$ satisfies all conditions of this example.

Example 8 (See also [1]).

Suppose that $g(s)$ is convex and non-negative for $s \geq 0$, concave and non-positive for $s \leq 0$, and satisfies

$$|g(s)| \leq c|s|^{\frac{p}{2}} \quad \forall s \in \mathbb{R} .$$

Then $(co_s f)(x,s,0) = f(x,s,0)$ but, in general, $f(x,s,z)$ is not convex in (s,z). Nevertheless it can be proved that there exists $N_o > 0$ such that for every $N \in]0,N_o]$ and for every $b \in L^p(\Omega)$

$$(co_{s,z} f)(x,s,0) = f(x,s,0) \qquad \forall x \in \Omega \qquad \forall s \in \mathbb{R} ,$$

thus $\psi(x,s) = f(x,s,0)$. This implies that $\lim\limits_{\varepsilon \to o_+} J_\varepsilon = J_o$ for every $N \in]0,N_o]$ and for every $b \in L^p(\Omega)$. The function $g(s) = s|s|^{\frac{p-2}{2}}$ satisfies all conditions of this example.

From the representation formula (b) we obtain also the following result.

Proposition 9 (See also [6]).

If g is decreasing, then $\psi(x,s) = (co_s f)(x,s,0)$ for a.a. $x \in \Omega$ and for all $s \in \mathbb{R}$.

Therefore also in the case of proposition 9 we have $\lim\limits_{\varepsilon \to o_+} J_\varepsilon = J_o$.

The first example where $\lim\limits_{\varepsilon \to o_+} J_\varepsilon \neq J_o$ was given by A. HARAUX and F. MURAT [7] for an optimal control problem with a constraint on the control v of the type

$$\alpha \leq v(x) \leq \beta \qquad \text{a.e. in } \Omega .$$

I shall now give two examples in which $\lim\limits_{\varepsilon \to o_+} J_\varepsilon \neq J_o$, without any constraint on the control v.

Example 10

Let $n = 1$, $p = 2$, $\Omega =]0,1[$, and let g be defined by

$$g(1) = \begin{cases} s & \text{if } s < 0 \\ \dfrac{s}{4} & \text{if } s \geq 0 . \end{cases}$$

Let $N > 6\pi^2 - 16$ and let $b \in L^2(\Omega)$. Then

$$\psi(x,s) < (co_s f)(x,s,0) = f(x,s,0)$$

for a.a. $x \in \Omega$ and for all $s > 0$. If in addition $b(x) > 0$ for every $x \in \Omega$, then

$$J_o = \inf_{u \in L^2(\Omega)} \int_\Omega f(x,u(x),0)dx > \inf_{u \in L^2(\Omega)} \int_\Omega \psi(x,u(x))dx = \lim_{\varepsilon \to o_+} J_\varepsilon .$$

Proof

The idea of the proof is to use the representation formula (c) of theorem 3 in order to majorize $\psi(x,s)$. By an easy change of variable we transform formula (c) into

$$\psi(x,s) = \inf \frac{1}{T} \int_o^T [N|u''(y) + g(u(y))|^2 + |u(y) - b(x)|^2] \, dy$$

where the infimum is taken over all $T > 0$ and over all functions $u \in H^2_{loc}(\mathbb{R})$ which are T-periodic and satisfy $\frac{1}{T} \int_o^T u(y)dy = s$.

All solutions of the equation

(3) $\quad u'' + g(u) = 0$

are periodic with period 3π . Moreover for every $s > 0$ there exists a solution u_s of the equation (3) such that

$$s = \frac{1}{3\pi} \int_o^{3\pi} u_s(y)dy .$$

This solution can be written explicitly :

$$u_s(y) = \begin{cases} \frac{s\pi}{2} \sin y & \text{if } -\pi \leqslant y \leqslant 0 \\ s\pi \sin \frac{y}{2} & \text{if } 0 \leqslant y \leqslant 2\pi . \end{cases}$$

Therefore if $s > 0$

$$\psi(x,s) \leqslant \frac{1}{3\pi} \int_o^{3\pi} |u_s(y) - b(x)|^2 \, dy = \frac{3\pi^2}{8} s^2 - 2b(x)s + |b(x)|^2 .$$

If $N > 6\pi^2 - 16$ we have

$$\frac{3\pi^2}{8} s^2 < (\frac{N}{16} + 1)s^2 .$$

Hence

$$\psi(x,s) \leqslant \frac{3\pi^2}{8} s^2 - 2b(x)s + |b(x)|^2 <$$

$$< (\frac{N}{16} + 1)s^2 - 2b(x)s + |b(x)|^2 = f(x,s,0)$$

for every $s > 0$. Moreover, if $b(x) > 0$ for every $x \in \Omega$, then

$$J_0 = \inf_{u \in L^2(\Omega)} \int_\Omega f(x,u(x),0)dx = (1 - \frac{16}{N+16}) \int_\Omega |b(x)|^2 dx >$$

$$> (1 - \frac{8}{3\pi^2}) \int_\Omega |b(x)|^2 dx =$$

$$= \inf_{u \in L^2(\Omega)} \int_\Omega [\frac{3\pi^2}{8} |u(x)|^2 - 2b(x)u(x) + |b(x)|^2] dx \geqslant$$

$$\geqslant \inf_{u \in L^2(\Omega)} \int_\Omega \psi(x,u(x))dx = \lim_{\varepsilon \to 0_+} J_\varepsilon .$$

In the following example the function g is a polynomial.

<u>Example 11</u>

Let $n = 1,\ p = 6,\ \Omega =]0,1[$,

$$g(s) = s^3 + s - \frac{5}{8} .$$

Then there exist $s_0 \in]0,1/2[$ and $K \in]0,+\infty[$ with the following property : if $b \in L^\infty(\Omega)$ and $N \geqslant K[1 + \|b\|^4_{L^\infty(\Omega)}]$, then

$$\psi(x,s_0) < (co_s f)(x,s_0,0) = f(x,s_0,0) \quad \text{for a.a. } x \in \Omega .$$

Moreover there exist $N_0 > 0$ such that for every $N \geqslant N_0$ there exists $b_N \in L^\infty(\Omega)$ for which

$$J_0 = \inf_{u \in L^2(\Omega)} \int_\Omega f(x,u(x),0)dx > \inf_{u \in L^2(\Omega)} \int_\Omega \psi(x,u(x))dx = \lim_{\varepsilon \to 0_+} J_\varepsilon .$$

The proof of example 11 is similar to the proof of example 10 (see [3], proposition 5.4). The idea is the same, but there are some technical difficulties due to the fact that the ordinary differential equation $u'' + g(u) = 0$ has no explicit solution in terms of elementary functions.

REFERENCES

[1] A. BENSOUSSAN : Un résultat de perturbations singulières pour systèmes distribués instables. C.R. Acad. Sci. Paris, Sér. I 296 (1983) 469-472.

[2] G. BUTTAZZO, G. DAL MASO : Γ-convergence et problèmes de perturbation singulière. C.R. Acad. Sc. Paris, Sér. I, 296 (1983) 649-651.

[3] G. BUTTAZZO, G. DAL MASO : Singular perturbation problems in the calculus of variations. To appear. On Ann. Scuola Norm. Sup. Pisa Cl.Sci.

[4] E. DE GIORGI, T. FRANZONI : Su un tipo di convergenza variazionale Rend. Sem. Mat. Brescia 3 (1979) 63-101.

[5] I. EKELAND, R. TEMAM : Convex analysis and variational problems. North-Holland, Amsterdam, 1976.

[6] A. HARAUX, F. MURAT : Perturbations singulières et problèmes de contrôle optimal : deux cas bien posés. C.R. Acad. Sci. Paris Sér. I, 297 (1983) 21-24.

[7] A. HARAUX, F. MURAT : Perturbations singulières et problèmes de contrôle optimal : un cas mal posé. C.R. Acad. Sci. Paris, Sér. I, 297 (1983) 93-96.

[8] A. HARAUX, F. MURAT : Influence of a singular perturbations on the infimum of some functionals. To appear.

[9] V. KOMORNIK : Perturbations singulières de système distribués instables. C.R. Acad. Sci. Paris, Sér. I, 296 (1983) 797-799.

Gianni DAL MASO
Università di Udine
Istituto di Matematica
Via Mantica, 3
33100 - UDINE (Italy)

B GAVEAU

Equations for hulls of holomorphy and foliations

This paper will treat two similar subjects : the construction of global hulls of holomorphy and the construction of plurisubharmonic measures. From the point of view of partial differential equations, these two subjects use degenerate semi-elliptic equations which are nonlinear and for which the traditional methods of solution do not apply. In both cases, we shall construct the solution of the equation without a priori estimates and fixed point theory, by a method very similar to the classical method of characteristics. We shall only sketch the methods; detailed articles have already been published. The first part is devoted to the construction of hulls of holomorphy in $\mathbb{C}^2$ and the construction of Levi flat hypersurfaces. The second part gives the definition of plurisubharmonic measures, their main properties and explicit examples. In all these constructions, we shall define and build a foliation naturally associated to our problem. The third part sketches possible generalizations in higher dimensions.

I - Construction of hulls of holomorphy in $\mathbb{C}^2$

1. K is a compact subset of $\mathbb{C}^n$, H(K) is the algebra of continuous functions on K which are uniform limits on K of holomorphic functions in the neighbourhood of K. Our problem is to study the spectrum of H(K) as a Banach algebra ; in particular, <u>can we construct a subset</u> $\hat{K} \subset \mathbb{C}^n$ <u>containing</u> K <u>such that all functions in</u> H(K) <u>can be analytically continued on a neighbourood of</u> $\hat{K}$?

2. The general method to construct $\hat{K}$ is as follows : we construct analytic disks with boundary in K, i.e. holomorphic mappings :

$$\varphi_t : \overline{\Delta(0,1)} \longrightarrow \mathbb{C}^n$$

with

$$\varphi_t(\partial\Delta) \subset K.$$

Here $\Delta(0,1)$ is the unit disks in $\mathbb{C}$ and t is a parameter in a connected set I such that if $t \longrightarrow t_o$, φ_t tends to a disk which is totally degenerate, i.e. φ_{t_o} is a constant mapping. It is well known that any holomorphic function near K can be holomorphically continued to a neighbourhood of

$$\bigcup_{t \in I} \varphi_t (\Delta(0,1)).$$

This construction gives good results when K is the closure of an open set in a strictly pseudo convex hypersurface in $\mathbb{C}$ but it is purely local and cannot take into account the global "multivalued phenomena" which imply that the hull spec H(K) is ramified over $\mathbb{C}^n$.

Moreover, when K is inside a submanifold of codimension > 1 in $\mathbb{C}^n$, the construction of analytic disk is very difficult (it was done for the first time by Bishop in [3]).

3. We shall now suppose that $K = \overline{U}$ where U is an open subset of an hypersurface S which is strictly pseudoconvex in $\mathbb{C}^2$.

Let us suppose that we can construct a hypersurface Σ with Levi form 0, intersecting the boundary of U in S, as in the diagram

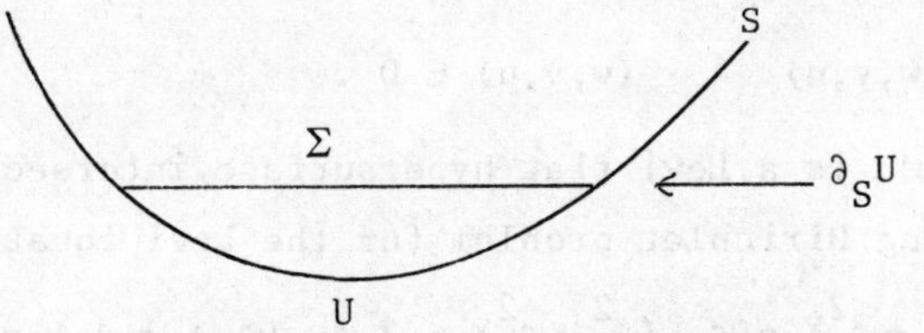

In [4], it was proved that holomorphic functions near $\overline{U}$ can be analytically continued to the domain between S and Σ.

Now, $\partial_S U$ is a compact surface of real dimension 2 in $\mathbb{C}^2$. The problem of the construction of the Levi flat hypersurface Σ leads us to the following question.

Let Γ be a compact surface in $\mathbb{C}^2$: find an hypersurface M such that

- $\Gamma \subset M$
- M has 0 Levi form.

Moreover is M unique ? and can we prove that the holomorphic functions near Γ extend holomorphically near M ?

Now, it is clear that the answer to these questions can be negative.

Namely, take

$$\Gamma = \{(z_1,z_2) \in \mathbb{C}^2 \ / \ |z_1| = |z_2| = 1\}.$$

There are many Levi flat hypersurfaces containing Γ $|z_1| = 1$, $|z_2| = 1$, $|z_1| = |z_2|^\alpha$ ... and Γ is holomorphically convex. One can also put, by an immersion, a two sphere in $\mathbb{C}^2$ such that it becomes knotted and Lagrangian ; then this sphere is totally real (i.e at any point of the sphere the tangent space is not complex), and so it is holomorphically convex. One can also imbed a two sphere in $\mathbb{C}^2$ such that it intertwines a divisor in $\mathbb{C}^2$, then its hull is ramified (see [8]).

4. In [5] and [2] we have found the hypothesis to give positive answers to the preceding questions. Let $z = x + iy$, $w = u + iv$ the coordinates in $\mathbb{C}^2$, D a domain in $\mathbb{R}^3$ given by

$$D = \{ (z,w) \in \mathbb{C}^2 \ / \ r(z,u) < 0 \ , \ v = 0 \}. \tag{1}$$

Let Γ be a graph over ∂D

$$\Gamma = \{(z,w) \in \mathbb{C}^2 \ / \ v = f(x,y,u) \ , \ (x,y,u) \in \partial D\}.$$

We look for M as a graph over D

$$v = \varphi(w,y,u) \qquad (w,y,u) \in D .$$

If we write that M is a Levi flat hypersurface intersecting Γ, we obtain the following Dirichlet problem for the Levi equation :

$$\begin{cases} (\varphi_{xx} + \varphi_{yy})(1+\varphi_u^2) + \varphi_{uu}(\varphi_x^2 + \varphi_y^2) - 2\varphi_{ux}(\varphi_u\varphi_x - \varphi_y) - \\ \qquad\qquad - 2\varphi_{uy}(\varphi_u\varphi_y + \varphi_x) = 0 \\ \varphi = f \quad \text{on} \quad \partial D . \end{cases} \tag{2}$$

This equation is quasi-linear semi-elliptic and the determinant of its principal symbol is identically 0 . Nevertheless we proved in [5] that

1) if the Dirichlet problem (2) has a solution, this solution is unique,

2) if ∂D is strictly pseudoconvex, we have an a priori estimate of the the first derivatives of the solution by the Levi form of ∂D and the first and second derivatives of the boundary data,

3) even if D is weakly pseudoconvex, the Dirichlet problem can be not well posed ; this is the case for

$$D = \{(z,u) \in \mathbb{R}^3 \,/\, |z| \leqslant 1 \,,\ 0 \leqslant u \leqslant 1\}$$

$$\varphi\,(z,u) = |z|^2 \quad \text{on } \partial D \,.$$

4) The Levi equation (2) does not come from a variational problem, even by multiplication by an integrating factor. But, multiplying by $\varphi_u(1+\varphi_u^2)^{-2}$, it can be put in divergence form.

5 . In [5], we have used a totally different method to solve (2) by using its geometrical interpretation ; (2) means that the hypersurface M is foliated in complex analytic subsets of dimension 1.

If we suppose that these sets intersect Γ, they induce a foliation by curves in Γ. But Γ is a surface and the Poincaré-Bendixson theory tells us what are the foliations of Γ. If Γ is a sphere or a surface of genus $g > 1$, the foliations have singularities. But we can say a priori where the singularities of this foliation are ; generically, the tangent space $T_p\Gamma$ is a purely real subspace of $\mathbb{C}^2$, but at some points (supposed isolated), $T_p\Gamma$ can become a complex subspace ; near these points, Γ can be written locally in the form

$$w = |z|^2 + \beta\,(x^2 - y^2) + O(|z|^3)$$

and p is said to be elliptic if $0 \leqslant \beta \leqslant 1$ and hyperbolic if $\beta > 1$. In [3], Bishop proved that the difference betwen the number of elliptic points and the number of hyperbolic is the Euler-Poincaré characteristic $\chi(\Gamma)$. But it is clear that the singularities of our foliation of Γ are exactly these points. So, we can predict where are the singularities of our (up to now unknown) foliation of Γ, just by looking at the way in which Γ is embedded in $\mathbb{C}^2$.

We shall limit ourselves to the following case :

1) Γ <u>is topologically a sphere and there are only two elliptic singular points</u> p_1, p_2 (and no other singular point),

2) ∂D <u>is strictly pseudoconvex (in the sense that</u> $\partial D \times \mathbb{R} \subset \mathbb{R}^3 \times \mathbb{R} \equiv \mathbb{C}^2$ <u>is strictly pseudoconvex)</u>

We then have

Theorem

If ∂D and f are C^{m+5}, there exists a unique hypersurface $\Phi \in C^{m,\alpha}(\overline{D} - \{p_1,p_2\}) \cap \text{Lip}\,\overline{D}$ for any $\alpha < 1$ which solves (2). Moreover the graph M of Φ over D is the hull of holomorphy of Γ.

6. Let us briefly sketch the proof of this theorem.

a) Using the results of [3], one can construct near p_1 and p_2 a one parameter family of complex analytic disks with boundaries covering a neighbourhood of $p_1 \cup p_2$ in Γ and we obtain in this way a small part of a Levi flat hypersurface near p_1 and p_2.

b) Then, starting from p_1, we begin this Bishop's construction as far as we can on a set T of real parameters t and we first prove that T is an open set : this means that if $\varphi_{t_o} : \Delta(0,1) \to \mathbb{C}^2$, $\varphi_{t_o}(\partial\Delta) \subset \Gamma$ is a disk, we have nearby disks with boundaries covering an open neighbourood of $\varphi_{t_o}(\partial\Delta)$. For that, let $F : \Delta(0,1) \to \mathbb{C}^2$ be a disk with $F(\partial\Delta) \subset \Gamma - \{p_1,p_2\}$.
We can define an index such that, if this index is 0, one can construct nearby disks near F by reducing Γ to a canonical form along a neighbourhood of $F(\partial\Delta)$. This index is the Maslov index when Γ is Lagrangian. Here, the index of $\varphi_{t_o}(\partial\Delta)$ is 0 because $\varphi_{t_o}(\partial\Delta)$ is obtained by a continuous deformation of small disks with 0 index.

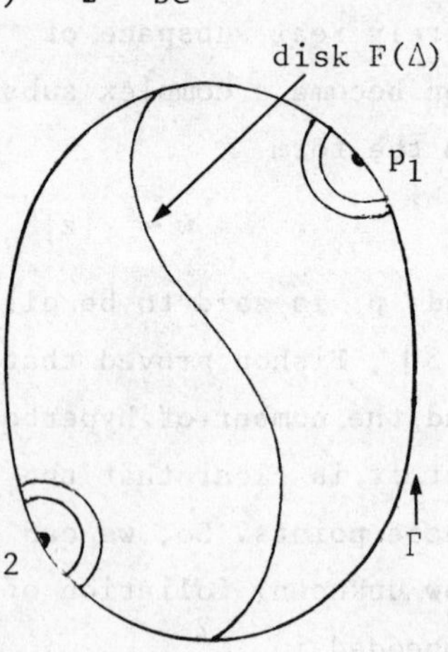

c) The third step is to prove regularity results ; if Γ is $C^{m,\alpha}$ and $f: \Delta(0,1) \to \mathbb{C}^2$ is a disk with $f(\partial\Delta) \subset \Gamma$ and $f|_{\partial\Delta} \in L^p_1(\partial\Delta)$ $(p > 1)$, then $f \in C^{m,\alpha}(\overline{\Delta})$ (in fact this result is a regularity result for a free boundary problem, namely for the holomorphic mapping $f : \Delta \to C^2$ with boundary in Γ).

d) We then prove that the set T is closed. Let t_n be a sequence in T converging to t_∞. We prove

α) an a priori estimate on the Lipschitz constant of $\varphi_{t_n}(\partial\Delta)$

β) that the disks $\varphi_{t_n}(\Delta)$ have no self intersection and can be projected on $\mathbb{R}^3$ in a one-to-one way.

γ) Using boundary value theory of conformal mappings, we can uniformize $\varphi_{t_n}(\Delta)$ by conformal transformations with a uniform $L^p_1(\partial\Delta)$ bound, and then, by step c), obtain $C^{m,\alpha}$ regularity.

e) Finally, we cover, starting from p_1, all $\Gamma-\{p_2\}$ by such disks. We must prove that the family that we have obtained near p_2 is the same as the one obtained by Bishop's construction near p_2 in step a). This is due to the following strange fact :
If two disks with boundaries in Γ in one-to-one projection on $\mathbb{R}^3$ have their boundaries touching at one point, then they are the same.

(This fact is false if the disks do not project in a 1.1 way on $\mathbb{R}^3$.)
[see [2] for details]

II - Construction of plurisubharmonic measures

1) We briefly recall the definitions and usefulness of this concept introtuced in [6]. Let D be a domain in $\mathbb{C}^n$, E a subset in ∂D and $P(E)$ the following class

(1) $$P(E) = \{u \text{ psh an } D/u \leqslant 1 \text{ an } D,\ \limsup_{z\to E} u \leqslant 0\}$$

and define

(2) $$\omega_{\complement E}(z) = \sup_{u\in P(E)} u(z).$$

This is the psh measure of $\complement E$. In one complex variable, this is the harmonic measure and it satisfies the Laplace equation. In several complex variables, $\omega_{\complement E}$ can be not usc in z, it is not a measure (σ-additive) in $\complement E$, but nevertheless we will keep the terminology "psh measure". It was easily proved in [6], that the $\omega_{\complement E}$ satisfies the complex Monge Ampère equation

(3) $$\det \partial\bar\partial\,\omega = 0$$

when it is a C^2 function.

2. The usefulness of ω_{CE} lies in its relation to the problem mentioned in I, namely the place where ω_{CE} is zero is the hull of holomorphy of E with respect to the class $A(\bar{D})$ of continuous functions on $\bar{D}$, which are holomorphic in D (in 1 complex variable, ω_{CE} does not vanish except in E). Moreover, ω_{CE} gives the best possible estimate of the growth of bounded holomorphic functions in D near ∂D:

$$|f(z)| \leqslant (\sup_E |f|)^{1-\omega_{CE}(z)} \times (\sup_D |f|)^{\omega_{CE}(z)} .$$

This is the generalization of the Hadamard inequality in $\mathbb{C}^n$. This also is a quantitative generalization of the concept of holomorphic hull.

3. If we want to find ω_{CE} we could say that we must solve the following Dirichlet problem:

$$\det \partial\bar{\partial}u = 0 \quad \text{in } D$$
$$u = 0 \quad \text{on } E \subset \partial D$$
$$u = 1 \quad \text{on } \partial D - E .$$

But the homogeneous complex Monge-Ampère equation is a degenerate elliptic equation which will propagate the singularities of boundary data. As we do not know anything about this equation, it seems important to construct many examples of explicit solutions, for example, when we have symmetry properties.

Here again, we shall not deal directly with equation (3) but we shall first give its geometrical interpretation. The equation $\det \partial\bar{\partial} u = 0$ means that the Levi form of u has at each point the eigen value 0 and that the kernels of $\partial\bar{\partial} u$ are an integrable distribution of complex spaces in one dimension, when the rank of $\partial\bar{\partial} u$ is $n-1$. This distribution can be integrated as a foliation $\mathscr{F}$ in analytic subsets of dimension one, and equation (3) tells us that u is harmonic along the leaves of the foliation $\mathscr{F}$ (as the leaves are one dimensional subsets in $\mathbb{C}^n$, it does not matter which metrics we use, they are all conformal on the leaves).

4. Let us now return to the situation described in the beginning of II and let us also suppose that there exists a group G of holomorphic mappings conserving D and E ; then, because ω_{CE} is a holomorphic invariant, G conserves ω_{CE} and the foliation $\mathscr{F}$ associated to ω_{CE}.

Our idea is as follows :

1) construct foliation $\mathscr{F}$ in complex analytic subsets of dimension one globally invariant by G . Unfortunately there are many foliations with this property and we shall have to guess which is the correct one.

2) When $\mathscr{F}$ is obtained, we construct u leaf by leaf by solving on each leaf L of $\mathscr{F}$ an ordinary Dirichlet problem in one complex variable, with boundary data 0 on $E \cap \overline{L}$ and 1 on $\underset{\partial D}{\complement} E \cap \overline{L}$.

3) It remains to check that this u is the correct function $\omega_{\complement E}$. By maximum principle on each leaf L , we see that

$$u \geqslant \omega_{\complement E} .$$

In the cases we shall treat, u will be C^2 by pieces, and we will directly prove that

a) u satisfies complex Monge-Ampère equation (3)

b) u is psh by computing the Euclidean Laplacian of u in $\mathbb{C}^n$.

This will prove that $u \leqslant \omega_{\complement E}$ (because u is in the class P(E)) and then $u = \omega_{\complement E}$.

4) To construct $\mathscr{F}$, we shall first construct a foliation in Levi flat hypersurfaces in $\mathbb{C}^2$; this will naturally induce $\mathscr{F}$.

Also a slight change in $\mathscr{F}$ can destroy either the psh property or the property of complex Monge-Ampère equation.

5. In [7] we have described two possible applications of this theory.

a) Let $D = \prod_{i=1}^{n} U_i$, $U_i \in \mathbb{C}$, $E_i \subset \partial U_i$

and let $\omega_{E_i}(z_i)$ be the harmonic measure of E_i in U_i .

Then the psh measure of $\prod_{i=1}^{n} E_i$ in $\prod_{i=1}^{n} U_i$ is

$$\sup \left(0 , \sum_{i=1}^{n} \omega_{E_i}(z_i) - n + 1\right)$$

and we can deduce that the holomorphically convex hull of $\complement \prod_{i=1}^{n} E_i$ for $A(\overline{D})$ is the set

$$\sum_{i=1}^{n} \omega_{E_i}(z_i) \leqslant n - 1 .$$

Here, the psh measure is not C^1 but only Lipschitzian. So we have absolutely no possibility of using the a priori estimate and traditional method to solve this kind of problem.

Let us also remark that the plurisubharmonic measure of $C\Pi E_i$ is

$$1 - \mathrm{Inf}(\omega_{E_1}(z_1),\dots, \omega_{E_n}(z_n)).$$

b) Let us also describe the following example : we start with a circled hypersurface in $\mathbb{C}^2$

$$(S) \qquad 2|z_2|^2 = \varphi(2|z_1|^2)$$

bounding a pseudoconvex D and we suppose also

$$\varphi(1) = 1$$

so that the bitorus

$$T^2 = \{(z_1,z_2) \ / \ |z_1| = |z_2| = (\sqrt{2})^{-1}\}$$

is in (S). Then the psh measure F of $C_S T^2$ is given by the following function:

1) $F = 0$ in the bidisk $\Delta^2 = \{(z_1,z_2) \ / \ |z_i| \leqslant (\sqrt{2})^{-1}\}$

2) $F(z_1,z_2) = \left[\log(2|z_1|^2)\right] \left[\log v\left(\dfrac{\log 2|z_2|^2}{\log 2|z_1|^2}\right)\right]^{-1}$

in $(D - \Delta^2) \cap \{|z_1| > (\sqrt{2})^{-1}\}$

where $v(x)$ is the function defined implicitly by

$$v^x = \varphi(v).$$

In this case the foliation $\mathscr{F}$ associated with F does not depend on φ, its leaves are given by

$$\sqrt{2}\, z_2 = (\sqrt{2}\, z_1)^c \ e^{i\theta}$$

where $c > 0$, $\theta \in \mathbb{R}$.

Here again the solution F is only Lipschitzian and not C^1. Other situations are described in [7] . Another example is treated by Bedford in [1].

III - Hulls of holomorphy in $\mathbb{C}^3$

1) In this paragraph, we want to give some general remarks about the construction of hulls of holomorphy in dimension > 2, and in fact we shall stay in dimension 3, i.e $\mathbb{C}^3$. We can again consider a strictly pseudoconvex hypersurface $S \subset \mathbb{C}^3$ and U a bounded open set in S; the problem is to find the spectrum of $H(\overline{U})$. Now $\partial_S U$ (boundary of U in S) is of real dimension 4 and so the situation is entirely different from what it was in $\mathbb{C}^2$ where $\partial_S U$ was of real dimension 2. The difference is that in $T_p(\partial_S U)$ there exists always a complex tangent line. (In $\mathbb{C}^2$, this happened only for some special points).

Now, the special points are those at which $T_p(\partial_S U)$ is a complex space of complex dimension 2 and, generically, this happens on a line in $\partial_S U$ of real dimension 1.

2) The natural generalization of Levi flat hypersurface in $\mathbb{C}^n$ is the following: a hypersurface Σ will be Levi flat if the determinant of the Levi form is identically 0. This implies that on an open set of Σ where the rank of the Levi form is $n - 2$, one can find foliation by analytic subsets of dimension 1.

As we shall see in a moment, the problem of finding a Levi flat hypersurface in $\mathbb{C}^3$ containing a given submanifold Γ_4 of real dimension 4, has an infinite number of solutions, and in fact the hull of holomorphy of Γ_4 will be, <u>in general</u>, a subset of $\mathbb{C}^3$ having interior point, and so containing open subsets in $\mathbb{C}^3$. We cannot say anything in general except on a special class of examples.

3) Write $\mathbb{C}^3 = \mathbb{C}^2 \times \mathbb{C}$ and call (z_1, z_2) and w the respective coordinates. Let D a domain in $\mathbb{C}^2$ and suppose that Γ_4 is given as

$$\Gamma_4 = \{(z_1, z_2, w) \in \mathbb{C}^3 / (z_1, z_2) \in \partial D \ , \ |w| = e^{-\varphi(z_1, z_2)}\}$$

where φ is a given function and we look for a Levi flat hypersurface Σ containing Γ_4 as given by

$$\Sigma = \{(z_1, z_2, w) \in \mathbb{C}^3 / (z_1, z_2) \in D \ , \ |w| = e^{-\Phi(z_1, z_2)}\}.$$

We find then that Φ satisfies the Dirichlet problem on D for the complex Monge-Ampère equation

$$(\star)\quad \begin{cases} \det\left(\dfrac{\partial^2\Phi}{\partial z_i\,\partial \bar z_j}\right)_{i,j=1,2} = 0 \text{ in } D \\ \Phi = \varphi \text{ in } \partial D. \end{cases}$$

Now this problem has many solutions; in fact usually we impose the condition that Φ should be psh in D and this gives uniqueness in this restricted class of functions.

But here, we have no reason to impose that Φ should be psh, and so the problem $(\star)$ is not well proved ; for example it can have a psh solution and a plurisuperharmonic solution and these two solutions are distinct in general (except if they are identical because φ is a boundary value of a pluriharmonic function on ∂D and so has a pluriharmonic extension Φ to D).

Call Φ^+ (sup Φ^-) the plurisuperharmonic (resp. the plurisubharmonic) solutions of problem $(\star)$ and suppose that D is strictly pseudoconvex : under this hypothesis, we know by Bremermann that Φ^+ and Φ^- exist and are unique in their respective class (provided φ is continuous) and we have also $\Phi^- \leqslant \Phi^+$ by maximum principle.

Theorem

<u>The hull of holomorphy of</u> Γ_4 <u>in</u> $\mathbb{C}^3$ <u>is the set</u> :

$$\hat\Gamma_4 = \{(z_1,z_2,w)/(z_1,z_2) \in \bar D \text{ , } e^{-\Phi^+(z_1,z_2)} \leqslant |w| \leqslant e^{-\Phi^-(z_1,z_2)}\}.$$

So, in general, $\hat\Gamma_4$ will contain an open subset of $\mathbb{C}^3$. We also remark that in this case, $T_p\,\Gamma_4$ contains only a complex space of dimension 1 at any point $p \in \Gamma_4$, which is the lift by the canonical projection $(z_1,z_2,w) \overset{\pi}{\to} (z_1,z_2)$ of the complex tangent line to ∂D at $\pi(p)$ and the other tangent direction $\frac{\partial}{\partial\theta}$ (θ argument if w) is conjugate to $\frac{\partial}{\partial|w|}$ which is not tangent to Γ_4.

4) We can also remark that in $\mathbb{C}^2$, if Γ_2 is given as a graph $|w| = e^{-\varphi(z)}$ $|w| = e^{-\varphi(z)}$, then the condition on $|w| = e^{-\Phi(z)}$ to be Levi flat, is just the Laplace equation $\frac{\partial^2\Phi}{\partial z\partial\bar z} = 0$.

So in some sense, the problem of finding in $\mathbb{C}^3$ an hypersurface with its Levi form of determinant 0 is similar to Monge-Ampère equation, and the problem of finding in $\mathbb{C}^2$ an hypersurface with Levi form 0, is similar to the Laplace equation. This gives an insight in the difficulty of the problem of hulls of holomorphy in $\mathbb{C}^3$.

BIBLIOGRAPHIE

[1] E. BEDFORD Michigan J. Math., 27, 1980, 365-370.

[2] E. BEDFORD, B. GAVEAU American J. of Math., 1983.

[3] E. BISHOP Duke Math. J. 32, 1965, 1-22.

[4] A. DEBIARD, B. GAVEAU J. Funct. Analysis, 48, 1976, 448-468.

[5] A. DEBIARD, B. GAVEAU Bull. Sci. Math., 1979.

[6] A. DEBIARD, B. GAVEAU dans Proc. conférence on analytic functions lecture note 798 (éd. J. Lawrynowicz).

[7] B. GAVEAU, J. KALINA Calculs explicites de mesures plurisousharmoniques et des feuilletages associés. Comptes Rendus, 1983, et Bull. Sci. Math. 1984.

[8] B. GAVEAU Cours au CIRME, Trento, Juillet 1982.

Bernard GAVEAU
Département de Mathématiques et
Equipe Associée au CNRS 761
Université de Paris VI
Place Jussieu - 75005 - PARIS

E GIUSTI

Quelques résultats de régularité pour les minima de fonctionnelles du calcul des variations

Le but de cet exposé est de présenter quelques résultats de régularité pour les minima de fonctionnelles régulières du calcul des variations

(1) $$F(u;\Omega) = \int_\Omega f(x,u,Du)\,dx .$$

Dans la formule (1), Ω est un ouvert borné de $\mathbb{R}^n$, $u : \Omega \to \mathbb{R}^N$ est une fonction de $W^{1,2}_{loc}(\Omega, \mathbb{R}^N)$ et f satisfait aux inégalités :

(2) $$|p|^2 - A \leqslant f(x,u,p) \leqslant A|p|^2 + 1).$$

Nous dirons que une fonction u de $W^{1,2}_{loc}(\Omega; \mathbb{R}^N)$ est un <u>minimum local</u> de F si pour toute $v \in W^{1,2}_{loc}(\Omega; \mathbb{R}^N)$ telle que $S = \operatorname{supp}(u-v) \subset\subset \Omega$ on a

$$F(u;S) \leqslant F(v;S).$$

Dans la suite nous traiterons surtout la régularité intérieure des minima locaux, en nous intéressant seulement marginalement aux problèmes au bord. En tout cas, nous nous bornerons au problème de Dirichlet. De plus, nous ne considèrerons seulement que le cas général $N>1$; c'est le cas où la variation première de la fonctionnelle donne lieu à un système d'équations différentielles. Les résultats que nous allons décrire seront naturellement moins forts que dans le cas scalaire, $N=1$, où l'on peut démontrer l'analogue du théorème de E. DE GIORGI, à savoir la continuité holdérienne des minima. Ici il n'est pas question en général de la régularité globale, et on n'arrivera qu'à donner des estimations de la dimension des singularités.

En vue de la recherche de théorèmes de plus en plus efficaces, nous serons amenés à spécialiser la forme de la fonction f. Un cas particulier mais de quelque intérêt est celui des fonctionnelles quadratiques, c'est à dire d'une fonction f de la forme

(3) $$f(x,u,Du) = \sum_{\alpha,\beta=1}^{n} \sum_{i,j=1}^{N} a_{ij}^{\alpha\beta}(x,u) D_\alpha u^i \, D_\beta \, u^j .$$

Pour ne pas compliquer l'écriture avec trop d'indices, nous écrirons dans ca cas

$$f(x,u,Du) = a(x,u)\, Du\cdot Du .$$

De plus nous traiterons en détail des fonctionnelles d'un type encore plus particulier, à savoir avec coefficients de la forme

(4) $$a_{ij}^{\alpha\beta}(x,u) = G_{ij}(x,u)\, g^{\alpha\beta}(x)$$

(coefficients séparés) dont l'interêt réside surtout dans leur intervention dans la théorie des applications harmoniques entre variétés riemaniennes, dont l'énergie en coordonnées locales prend la forme (3), (4).

Sauf mention contraîre, les résultats exposés ont été obtenus en coopération avec M. GIAQUINTA.

1. Régularité partielle

Les premiers résultats de régularité pour des fonctionnelles générales (1), (2) ont été démontrés dans [5] :

Théorème 1 - Dans les seules hypothèses (2), tout minimum local u de F est dans $W_{loc}^{1,p}(\Omega, \mathbb{R}^N)$, pour un certain $p > 2$. De plus,

(5) $$\left\{ ⨍_{B(x_o,R)} (1+|Du|^p)\, dx \right\}^{1/p} \leqslant c \left\{ ⨍_{B(x_o,2R)} (1+|Du|^2)\, dx \right\}^{1/2}$$

pour toute boule $B(x_o,2R) \subset\subset \Omega$, $⨍$ dénotant la moyenne intégrale :

$$⨍_A = \frac{1}{\text{mes}(A)} \int_A .$$

Le théorème 1 n'est qu'une étape intermédiaire pour les résultats qui suivent. Dans le cas $n = 2$ il implique d'ailleurs la continuité de la fonction u, grâce au théorème d'immersion de SOBOLEV.

Pour arriver à montrer la régularité partielle il faut ajouter des hypothèses sur la fonction f. Plus précisément :

(i) Pour tout $(x,u) \in \Omega \times \mathbb{R}^N$, $f(x,u,.)$ est deux fois différentiable, et l'on a

(6) $\quad |f_{pp}(x,u,p)| \leqslant L$

(7) $\quad f_{p^i_\alpha p^j_\beta}(x,u,p)\, \xi^i_\alpha \xi^j_\beta \geqslant \nu|\xi|^2 \,, \quad \nu > 0$

(on somme ici et par la suite sur les indices répétés, les indices grecs variant de 1 à n, les latins de 1 à N).

(ii) Pour tout $p \in \mathbb{R}^{nN}$ la fonction $(1+|p|^2)^{-1} f(x,u,p)$ est höldérienne en (x,u), uniformément par rapport à p.

Sous ces hypothèses l'on démontre la régularité partielle de la solution, c'est à dire que pour tout minimum local u de F il existe un ouvert $\Omega_o \subset \Omega$ (dépendant de u) tel que $\Omega - \Omega_o$ à mesure nulle, et que u soit de classe $C^{1,\alpha}(\Omega_o)$ pour un certain $\alpha > 0$([6], [12]).

L'ensemble singulier $\Sigma = \Omega - \Omega_o$ est fermé et en général non vide. Les exemples qui suivent donnent des cas qui sont pratiquement les seuls qu'on connaisse, où $\Sigma = \emptyset$.

<u>Exemple 1</u> - (Fonctionnelles qui ne dépendent que du module du gradient) Nous supposerons que $f = f(p)$ satisfait la condition (i) ci-dessus, et qu'elle soit de la forme

$$f(p) = g(|p|^2).$$

Dans ce cas tout minimum, mais aussi tout point stationnaire, de F est une fonction de classe $C^{1,\alpha}(\Omega)$ ([4], [20]).

<u>Exemple 2</u> - (Perturbations d'une fonctionnelle quadratique)
Ici f est de la forme

$$f(x,u,Du) = A(x)\, Du \cdot Du + g(x,u,Du)$$

avec des coefficients A(x) qui ne dépendent pas de u, qui sont continus et satisfont la condition de Legendre-Hadamard

$$A^{\alpha\beta}_{ij}(x)\xi_\alpha \xi_\beta \eta^i \eta^j \geqslant \nu|\xi|^2 \,|\eta|^2 \quad ; \quad \nu > 0$$

tandis que g est une fonction à croissance lente :

$$|g(x,u,p)| \leqslant L(1+|p|^r) \quad ; \quad r < 2 \,.$$

Dans ce cas tout minimum local de F est une fonction Holder-continue pour tout exposant $\alpha < 1$([6]). Si en plus g ne dépend pas de p, et vérifie

$$|g(x,u) - g(x,v)| \leqslant L|u-v|^\gamma \quad ,$$

et si les coefficients A sont des fonctions σ-Hölder-continues, tout minimum de F est une fonction de $C^{1,\alpha}(\Omega)$ ([6],[8]) avec

$$\alpha = \min\{\sigma, \frac{\gamma}{2-\gamma}\} .$$

L'exposant α ne peut pas être amélioré, même si $N = 1$ dans le cas où

$$f(x,u,Du) = |Du|^2 + 2u$$

(voir [15]).

Exemple 3 - (Stabilité des solutions régulières). Si tout minimum local de

$$F_o(u;\Omega) = \int_\Omega a_o(x,u) Du\cdot Du \, dx$$

est régulier (par exemple $C^{o,\alpha}$), et si a_k converge vers a_o, alors toute suite u_k de minima de F_k est définitivement régulière. Plus précisement supposons que les coefficients a_o soient bornés et elliptiques :

(5) $$a_o(x,u)\ \xi\cdot\xi \geqslant \nu|\xi|^2 \quad , \quad \nu > 0 ,$$

et considérons une suite a_k qui converge vers a_o uniformément sur tout compact de $\Omega \times \mathbb{R}^N$. Pour tout $k \in \mathbb{N}$, soit u_k un minimum de F_k, et supposons que u_k converge dans L^2_{loc}-faible vers une fonction u_o, qui sera nécessairement un minimum local de F_o.

Soit maintenant x_o un point régulier pour u_o. Alors toute u_k, pour k suffisamment grand, sera régulière dans un voisinage de x_o (qui ne dépend pas de k).

Par conséquent, si u_o n'a pas de points singuliers dans un compact $K \subset \Omega$, il en sera de même pour les u_k, à partir d'un indice k_o assez grand. En particulier, si F_o n'a que des solutions régulières, les points singuliers de u_k devront nécessairement s'approcher de la frontière.

Si l'on se borne à des fonctionnelles quadratiques (3), il es possible d'améliorer l'estimation de la dimension de l'ensemble des singularités.

Théorème 2 [5] *Soit F une fonctionnelle quadratique (3), avec des coefficients $a(x,u)$ bornés, continus et satisfaisants à la condition d'ellipticité (5). Alors tout minimum local u de F est une fonction Holdérienne dans un ouvert $\Omega_o \subset \Omega$ (dépendant de u) et l'on a pour un certain $q > 2$:*

$$H_{n-q}(\Omega - \Omega_o) = 0$$

H_s dénotant la mesure de Hausdorff s-dimensionnelle.

2. Fonctionnelles quadratiques à coefficients séparés

Le théorème précédent donne le résultat le plus général pour des fonctionnelles quadratiques. Si l'on veut améliorer la borne supérieure pour la dimension de l'ensemble singulier Σ, on sera obligé de spécialiser à nouveau les coefficients.

Le point de départ est constitué du théorème 2. Une idée ingénue de la dimension d'une ensemble suggère que, comme la dimension Σ est strictement plus petite que $n-2$, elle ne saurait dépasser $n-3$.

Malheureusement, il s'agit ici de la dimension de Hausdorff, qui est définie à partir de la mesure :

$$\dim(\Sigma) = \inf \{ s : H_s(\Sigma) = 0 \} ,$$

et qui peut prendre en principe toute valeur réelle entre zéro et n.

Néanmoins, l'idée que la dimension doit tomber à $n-3$ n'est pas dépourvue d'un fond de vérité, du moins si les coefficients sont de la forme (4). La démonstration repose sur une notion de solution tangente, analogue à celle de cône tangent introduite par E. DE GIORGI dans l'étude des surfaces minima (voir e.g. [10]). Pour l'illustrer sans être encombré de détails techniques on se bornera au cas $n = 3$.

Supposons alors qu'une fonction bornée u, qui minimise la fonctionnelle F, aie une suite de points singuliers x_k qui convergent vers 0. Si l'on pose $R_k = 2|x_k|$, et

$$u_k(x) = u(R_k x) ,$$

la fonction u_k est un minimum de la fonctionnelle quadratique F_k, dont les coefficients sont

$$a_k(x,u) = a(R_k x,u).$$

De plus, u_k a un point singulier (en dehors de l'origine) y_k, avec $|y_k| = 1/2$.

Les u_k étant uniformément bornées, on peut supposer qu'elles convergent dans L^2-faible à une fonction v. Il n'est pas difficile de démontrer que

v minimise la fonctionnelle

$$F_o(v;\Omega) = \int_\Omega a(0,v)Dv\cdot Dv\,dx .$$

C'est à ce point qu'on fait intervenir la forme particulière (4) des coefficients. Car, si l'on a

$$a_{ij}^{\alpha\beta}(x,u) = G_{ij}(x,u)\, g^{\alpha\beta}(x)$$

il est possible de démontrer un résultat de monotonie de la fonctionnelle $F(u;B_r)$ par rapport au rayon r. Ce résultat permet alors de conclure que <u>la fonction limite</u> v <u>est homogène de degré zéro</u>.

D'autre part l'on peut supposer que les points singuliers y_k (qui sont tous sur la sphère de rayon 1/2) convergent vers un point y_o de la même sphère. D'après l'exemple 3, y_o doit nécessairement être un point singulier de v.

Mais alors la droite qui passe par 0 et y_o serait toute formée de points singuliers (à cause de l'homogénéité de v) et donc l'ensemble singulier Σ aurait dimension au moins 1. Cela contredit le théorème 2. Nous avonc donc démontré :

<u>Théorème 3 - [7] Si</u> $n=3$ <u>et si les coefficients</u> $a_{ij}^{\alpha\beta}$ <u>ont la forme (4), alors l'ensemble singulier de tout minimum</u> u <u>de la fonctionnelle quadratique</u> F <u>est formé au plus de points isolés.</u>

Quand $n > 3$, le raisonnement précédent ne marche plus, et il faut l'améliorer quelque peu. L'idée est encore suggérée par les surfaces minimales (voir [3] où [10]). Supposons que $H_k(\Sigma) > 0$, pour un certain $k > 0$ et que $0 \in \Sigma$ soit un point de densité k-dimensionnelle positive pour Σ. En procédant comme auparavant on parvient à une fonction v homogène de degré zéro, qui minimise la fonctionnelle F_o, et pour laquelle on a encore $H_k(\Sigma_v) > 0$.

Soit maintenant $x_o \neq 0$ un point de densité k-dimensionnelle positive pour Σ_v. Sans limiter la généralité on peut supposer que $x_o = (0,0,\dots,0,1)$. Nous pourrons faire exploser encore la fonction v autour de x_o en considérant la suite

$$v_j(x) = v(x_o + j^{-1}x).$$

Une sous suite de v_j convergera vers une fonction w, qui sera encore un minimum de F_o. A cause de l'homogénéité de v, la fonction w sera indépendante de x_n .

Il n'est pas difficile maintenant de voir que w , considérée comme une fonction de n-1 variables, minimise F_o dans $\mathbb{R}^{n-1}$, et que ses points singuliers ont une mesure (k-1)-dimensionnelle positive en $\mathbb{R}^{n-1}$. On peut alors en déduire deux conséquences :

(I) Considérons une fonctionnelle quadratique à coefficients séparés qui ne dépendent pas de x , et supposons que ses minima n'aient pas de singularités en dimension h . Alors les singularités en dimension $h+1$ sont au plus des points isolés.

(II) En décroissant successivement la dimension de l'espace et de l'ensemble Σ , on peut conclure que la dimension des singularités en dimension n est au plus $n-h-1$ [7] . En particulier, grâce au théorème 2, la dimension des singularités d'une fonctionnelle quadratique arbitraire à coefficients séparés ne dépasse pas $n-3$.

3. Applications harmoniques minimisantes

Un cas d'un certain intérêt est celui des applications harmoniques entre deux variétés riemaniennes. Si m^n et M^N sont de telles variétés, de dimension n et N respectivement, et si $u : m \to M$, ont peut définir l'énergie de u comme la quantité

$$E(u) = \int_m |du|^2 .$$

En coordonnées locales :

$$E(u) = \int_m G_{ij}(u)\, g^{\alpha\beta}(x)\, D_\alpha u^i\, D_\beta u^j \sqrt{g}\, dx \tag{6}$$

où $g_{\alpha\beta}$ et G_{ij} sont les tenseurs métriques de m et M respectivement, $g^{\alpha\beta}$ est l'inverse de $g_{\alpha\beta}$ et $g = \det(g_{\alpha\beta})$.

Une application harmonique est un point stationnaire de E ; nous serons intéressés aux applications qui minimisent localement l'énergie.

Il est aisé de voir que la fonctionnelle (6) est à coefficients séparés. D'autre part les résultats de régularité du théorème 3 ne sauraient pas s'appliquer directement à cause de l'impossibilité de se servir globalement de coordonnées locales. Il est clair, puisque la régularité est une propriété locale, qu'on peut se borner à un voisinage d'un point x_o de m ; de plus, à cause de l'exemple 3 , l'on peut aussi se réduire au cas où m est un disque plat D^n .

D'autre part, il n'est pas ainsi pour la variété M , la localisation n'étant possible en principe que si u est continue (un cas sans intérêt, car c'est exactement la continuité de u qui est en cause), ou si M est couverte par une seule carte, c'est à dire si M est $\mathbb{R}^N$ munis d'une métrique riemannienne arbitraire. Dans ce cas l'on peut effectivement conclure que toute application harmonique minimisante vérifie $\dim \Sigma \leqslant n-3$.

Il faut remarquer que cette restriction sur M n'est pas nécessaire, car Schoen et Uhlenbeck [17] ont démontré indépendemment le même résultat pour des variétés quelconques.

Il est aussi à remarquer que si on considère le problème de Dirichlet, c'est à dire si l'on s'occupe des minima de E sous la condition que $u = \varphi$ au bord de m , l'ensemble singulier Σ ne saurait pas arriver jusqu'au bord comme l'on démontré indépendemment encore, JOST et MEIER [14] dans le cadre ci-dessus et SCHOEN et UHLENBECK [18] pour les applications harmoniques.

Je ne connais pas d'exemples d'applications harmoniques minimisantes avec un ensemble singulier de dimension $n-3$. En général, on peut penser que la dimension, et même l'existence des singularités dépendra de la géométrie de la variété cible M^N , et non pas de celle de départ m^n. Comme on l'a déja remarqué, on peut se borner à un disque D^n.

Les premiers résultat généraux d'existence (et de régularité) d'applications harmoniques sont dus à EELLS et SAMPSON [2] ; ils démontrent que si la courbure sectionnelle de M est non-positive, toute classe d'homotopie a un représentant harmonique. Quoique très intéressant, ce résultat se pose dans un cadre différent du nôtre, car (i) il fait des hypothèses très strictes sur la variété M , et (ii) les applications ainsi trouvées ne sont pas des minima de l'énergie.

Un résultat plus proche de notre point de vue est dû à HILDEBRANDT, KAUL et WIDMAN [1] . Ce résultat montre entre outre la régularité de toute application harmonique (même non minimisante) du disque D^n dans M^N, pourvu que son image soit contenue dans une boule géodésique $\mathscr{B}_R(q)$ qui ne rencontre pas le <u>cut-locus</u> de q , et de rayon R assez petit :

$$R < \pi/2\sqrt{\kappa}$$

où κ est une borne supérieure pour la courbure sectionnelle de M^N. Il est à remarquer que si M est la sphère unitaire S^N ($\kappa=1$) et si $R = \pi/2$, la boule $\mathscr{B}_R$ est une hémisphère. On est donc amené à étudier les applications harmoniques de D^n dans S^N, dont l'image soit contenue dans un hémisphère, un cas qui échappe au théorème de Hildebrandt, Kaul et Widman.

Il y a aussi une autre raison pour cette étude, notamment la relation entre les applications harmoniques sur la sphère et les surfaces minimales. Nous avons vu que les méthodes des surfaces minimales pouvaient être appliquées avec succès à l'étude des fonctionnelles quadratiques, et en particulier aux applications harmoniques. En effet il y a un lien très étroit entre les deux problèmes, car RUH et VILMS [16] ont montré que si m est une hypersurface minimale de dimension n, son <u>Gauss map</u> u est une application harmonique de m dans S^n. Si en plus la surface m est un graphe, l'image de u est contenue dans l'hémisphère supérieur

$$S^n_+ = \{ y \in \mathbb{R}^{n+1} : |y| = 1 , y_{n+1} > 0 \} .$$

C'est le cas ici du problème de Bernstein général, tandis que si la fonction f (dont m est le graphe) a un grandient borné dans $\mathbb{R}^n$, l'image de son <u>Gauss map</u> est contenue dans quelque $\mathscr{B}_R$ avec $R < \pi/2$, pour lequel le théorème de Hildebrandt, Kaul et Widman serait applicable.

Pour les applications minimisantes dans un hémisphère S^N_+ il y a des résultats très récents, qui confirment l'analogie avec les surfaces minimales.

D'abord, HILDEBRANDT, KAUL et WIDMAN avaient déja remarqué que l'application équatoriale

$$U(x) = x/|x|$$

de D^n dans S^n est un point critique pour l'énergie. Plus récemment, JÄGER et KAUL [13] ont démontré que U est un minimum pour l'énergie (naturellement entre les applications qui ont les mêmes valeurs au bord) si $n \geq 7$, mais elle n'est même pas stable si $3 \leq n \leq 6$. BALDES [1] a considéré l'application équatoriale U de D^n dans un ellipsoïde de révolution :

$$Q^n = \{ y \in \mathbb{R}^{n+1} : y_1^2 + \dots + y_n^2 + y_{n+1}^2/a^2 = 1 \}$$

et a démontré que U est strictement stable si $a^2 > 4(n-1)/(n-2)^2$, et instable si $a^2 < 4(n-1)/(n-2)^2$.

La question a été reprise par GIAQUINTA et SOUCEK [9], qui ont démontré les théorèmes suivants (j'ai vu aussi mentionné dans [1] un article de SCHOEN et UHLENBECK [19], apparemment sur le même sujet, mais je n'ai pas encore réussi à les voir).

<u>Théorème 4.</u> <u>Si</u> $3 \leqslant n \leqslant 6$, <u>toute application harmonique minimisante de</u> D^n <u>dans</u> S_+^N <u>est régulière.</u>

Il suit de la remarque (I) que les points singuliers d'une application minimisante de D^7 dans S_+^N sont isolés.

De plus l'analogue du théorème de Bernstein est encore vrai.

<u>Théorème 6.</u> <u>Si</u> $n \leqslant 6$, <u>toute application harmonique minimisante de</u> $\mathbb{R}^n$ <u>dans</u> S_+^N <u>est constante.</u>

BIBLIOGRAPHIE

[1] BALDES A. Stability and Uniqueness Properties of the Equator map from a Ball into and Ellipsoïd. Prepint 1983.

[2] EELLS J. et SAMPSON J.H. - Harmonic maps of Riemannian manifolds. Amer. J. of Math. 86 (1964) 109-160.

[3] FEDERER H. The singular set of area minimizing rectifiable currents with codimension one and of area minimizing flat chains modulo two with arbitrary codimension. Bull. Am. Math. Soc. 76 (1970) 767-771.

[4] GIAQUINTA M. - Multiple integrales in the calculus of variations and nonlinear elliptic systems. Annals of Math. Studies 105 (1983).

[5] GIAQUINTA M. et GIUSTI E. - On the regularity of the minima of variational integrals. Acta Math. 148 (1982) 31-46.

[6] GIAQUINTA M. et GIUSTI E. - Differentiability of the minima of nondifferentiable functionnals. Inv. Math. 72 (1983) 285-298.

[7] GIAQUINTA M. et GIUSTI E. - The singular set of minima of certain quadratic functionnals. Preprint 1981.

[8] GIAQUINTA M. et GIUSTI E. - Sharp estimates for the derivatives of local minima of variational integrals. Prepint 1983.

[9] GIAQUINTA M. et SOUČEK J. - Harmonic maps into a hemisphere. Prepint 1983.

[10] GIUSTI E. Minimal surfaces and functions of bounded variation. Notes on Pure Math. Canberra 1977.

[11] HILDEBRANDT S. KAULT H. et WIDMAN K-O - An existence theorem for harmonic mappings of Riemannian manifolds. Acta Math 138 (1977) 1-16.

[12] IVERT P.A. Partial regularity of vector valued functions minimizing variational integrals. Prepint 1982.

[13] JAGER W. et KAUL H. - Rotationally symmetric harmonic maps from a ball into a spere and the regularity problem for weak solutions of elliptic systems. Prepint 1983.

[14] JOST J. et MEIER M. - Boundary regularity for minima of certain quadratic functionnals. Math. Ann. 262 (1983) 549-561.

[15] PHILLIPS D. A minimization problem and the regularity of solutions in presence of a free boundary. Prepint 1982.

[16] RUH E.A. et VILMS J. - The tension field of the Gauss map. Trans. Am. Math. Soc. 149 (1970) 569-573.

[17] SCHOEN R. et UHLENBECK K. - A regularity theory for harmonic maps. J. Diff. Geom. 17 (1982) 307-335.

[18] SCHOEN R. et UHLENBECK K. - Boundary regularity and the Dirichlet problem for harmonic maps. J. Diff. Geom. 18 (1983) 253-268.

[19] SCHOEN R. et UHLENBECK K. - Regularity of minimizing harmonic maps into the sphere. Preprint 1983.

Enrico GIUSTI
Departamento di Matematica
Universita di Fiorenza
FLORENCE (Italie)

J LERAY

Divers prolongements analytiques de la solution du problème de Cauchy linéaire

En employant une précision que j'avais jadis apportée au théorème de Cauchy-Kowaleski [L.1] , Y. HAMADA et A. TAKEUCHI, d'abord seuls [H.T.] , puis avec ma collaboration [H.L.T.,1] ont récemment effectué des prolongements analytiques de la solution du problème de Cauchy linéaire. En perfectionnant nos procédés, j'ai obtenu un résultat, dont j'ai dans [L,2] donné l'énoncé et esquissé la preuve. Je donne ici, sans preuve, des conséquences aisées de ce résultat, qui l'englobent et montrent sa portée. Un exposé détaillé de ces recherches paraîtra prochainement [H.L.T.,2] .

1. Le problème étudié

Soit Ω une variété analytique complexe de dimension complexe n . Notons ω un point arbitraire de Ω et $x' = (x_1,\ldots,x_n)$ des coordonnées analytiques locales de ω . Supposons Ω connexe, paracompact, non compact. Soit $\overline{\Omega}$ le compactifié de Ω par adjonction d'un point $\partial\Omega$, appelé "point à l'infini" ; les voisinages ouverts de $\partial\Omega$ sont les complémentaires dans $\overline{\Omega}$ des parties compactes de Ω .

Soit Σ une surface de Riemann, non compacte et simplement connexe. Notons σ un point arbitraire de Σ et x_o une coordonnée analytique locale de σ . Un point α de Σ est donné.

Notons

$$x = (\sigma,\omega) \in X = \Sigma \times \Omega \quad ;$$

x a donc les coordonnées locales $(x_o,x') = (x_o,x_1,\ldots,x_n)$.

Notons

$$\alpha \times \Omega = \{(\sigma,\omega) \in \Sigma \times \Omega \; ; \; \sigma = \alpha\} .$$

Soit a un opérateur différentiel d'ordre m, holomorphe dans X au voisinage de $\alpha \times \Omega$, opérant sur les fonctions numériques holomorphes. Nous supposons qu'aucune hypersurface $\sigma \times \Omega$ n'est, en aucun de ses points, caractéristique pour l'opérateur a.

Soient v et w deux fonctions numériques holomorphes au voisinage de $\alpha \times \Omega$. Nous étudions le problème de Cauchy : trouver une fonction numérique, holomorphe, u, telle que :

(1) $au = v$; $u - w$ s'annule m fois sur $a \times \Omega$.

Notre but est d'expliciter, aussi simplement que possible, des voisinages de $\alpha \times \Omega$, aussi grands que possible, sur lesquels le problème (1) possède une solution.

2. Construction de métriques riemanniennes sur Σ.

Notons l'expression locale de l'opérateur a :

(2.1) $$\sum_{|\lambda|\leq m} a_\lambda(x)\, D^\lambda , \quad \text{où } \lambda = (\lambda_o, \lambda_1, \ldots, \lambda_n) \in \mathbb{N}^{n+1}, \ |\lambda| = \lambda_o + \ldots + \lambda_n ,$$

$$D^\lambda = D_o^{\lambda o} \ldots D_n^{\lambda n}, \quad D_k = \partial/\partial x_k .$$

Notons $T_x^\star(X)$ le fibré cotangent de X et $T_x^\star(X)$ sa fibre de projection $x \in X$. On a :

$$T_x^\star(X) \ni \xi = (\xi_o, \xi') , \text{ où } \xi_o \in T_\sigma^\star(\Sigma) , \ \xi' \in T_\omega^\star(\Omega).$$

l'expression locale du polynôme caractéristique de a est :

(2.2) $$g(x;\xi) = \sum_{|\lambda| = m} a_\lambda(x)\, \xi^\lambda .$$

On a :

(2.3) $$g(x;\xi) = \sum_{r=o}^{m} g_r(x_o, \omega; \xi')\, \xi_o^r ,$$

g_r étant un polynôme en ξ', homogène de degré $m - r$; il dépend du choix de la coordonnée locale x_o. Par hypothèse l'hypersurface $\sigma \times \Omega$ n'est caractéristique en aucun de ses points, c'est à dire :

(2.4) $$(\forall\, x_o, \omega) \qquad g_m(x_o, \omega) \neq 0 .$$

L'équation caractéristique est l'équation d'inconnue ξ_o :

$$\sum_{r=0}^{m} g_r(x_o,\omega;\xi)\ \xi_o^r = 0 \ ;$$

ses racines, appelées racines caractéristiques, servent à l'étude de la propagation des singularités, quand les données présentent des singularités, ce qui n'est pas le cas ici. Nous emploierons la majorante $\rho(x_o,\omega;\xi')$ du module des racines caractéristiques définie comme suit :

Si $g_r(x_o,\omega;\xi') = 0$ en $(x_o,\omega;\xi')$ pour $r = 0,\ldots,m-1$, alors

$$\rho(x_o,\omega;\xi') = 0 \ .$$

Sinon $\rho(x_o,\omega;\xi')$ est l'unique racine $\rho > 0$ de l'équation

(2.5) $$\sum_{r=o}^{m-1} |g_r(x_o,\omega;\xi')|\ \rho^r = |g_m(x_o,\omega)|\ \rho^m \ .$$

La fonction $\rho(.,.;.)$ est continue.

(2.6) $$(\forall\ \theta \in \mathbb{C}) \quad \rho(x_o,\omega;\theta\xi') = |\theta| \ . \ \rho(x_o,\omega;\xi') \ .$$

La fonction $x_o \mapsto \log\ \rho(x_o,\omega;\xi')$ est sous-harmonique ou identique à $-\infty$.

L'expression $\rho(x_o,\omega;\xi')\ |dx_o|$ est indépendante du choix de la coordonnée locale x_o de σ et du choix des coordonnées locales de Ω.

Nous dirons qu'une partie fermée Φ de $T^\star(\Omega)$ est admissible quand elle possède les deux propriétés suivantes :

(2.7) la restriction à Φ de la projection canonique $T^\star(\Omega)$ est propre, c'est à dire : l'ensemble des points de Φ dont la projection canonique appartient à un compact arbitraire de Ω est un ensemble compact ;

la fonction

(2.8) $$x_o \mapsto \rho_{x_o} = \sup_{(\omega;\xi')\in\Phi} \rho(x_o,\omega;\xi')$$ est localement bornée.

Alors :

la fonction $x_o \mapsto \log\ \rho_{x_o}$

est sous-harmonique ou identique à $-\infty$.

L'expression

(2.9) $$ds = \rho_{x_o} \ |dx_o|$$

est indépendante du choix de la coordonnée locale x_o .

Par suite, ds^2 <u>est riemannien, conforme à la structure analytique complexe de</u> Σ <u>et à courbure</u> $\leqslant 0$ sur la partie de Σ où $\rho_{x_o} \neq 0$.

Nous définissons sur Σ

(2.10) $$\text{dist}(\sigma,\alpha) = \inf \int_\alpha^\sigma ds .$$

3. Un premier énoncé

Etant donné une fonction de classe C^1

$$R : \Omega \to \mathbb{R}$$

notons

$$\nabla R(\omega) = (R_{x_1},\ldots,R_{x_n}) \text{ l'élément de } T^\star_\omega(\Omega) \text{ tel que}$$

(3.1) $$dR(\omega) = \text{Re} \ (\sum_{k=1}^{n} R_{x_k} \ dx_k).$$

Puisque Ω est paracompact, nous pouvons choisir une fonction de classe C^1

$$R : \Omega \to \dot{\mathbb{R}}_+ \text{ telle que } \lim_{\omega\to\partial\Omega} R(\omega) = \infty .$$

Nous pouvons alors choisir une fonction continue

$$B[.] : \dot{\mathbb{R}}_+ \to \dot{\mathbb{R}}_+ = \ ]0,\infty[$$

telle que

(3.2) $$\Phi = \{(\omega, \frac{\nabla R(\omega)}{B[R(\omega)]}) \in T^\star(\Omega) \ ; \ \omega \in \Omega \}$$

soit admissible. Définissons :

(3.3 $$F[R] = \int_R^\infty \frac{dR'}{B[R']} \leqslant \infty \ , \ F(\omega) = F[R(\omega)] .$$

PROPOSITION I.

<u>Sous les conditions précédentes, le problème (1) possède une unique solution holomorphe sur le domaine</u>

(3.4) $$\Delta = \{(\sigma,\omega) \in \Sigma \times \Omega \ ; \ \text{dist.} \ (\sigma,\alpha) < F(\omega) \} ,$$

<u>si les données</u> a <u>et</u> v <u>du problème</u> (1) <u>sont holomorphes sur ce domaine</u> Δ .

Note

Si $F = \infty$, c'est-à-dire si l'intégrale (3.3) diverge, alors $\Delta = \Sigma \times \Omega$.

Note

Quand on peut choisir $B[R] = R^{-2}$, c'est-à-dire $F(\omega) = 1/R(\omega)$, alors la proposition I est identique au théorème dont [L2] a donné l'énoncé et esquissé la preuve.

Exemple I

Supposons que Ω est $\mathbb{C}^n$ muni d'une structure hermitienne et que, pour $r = 0,\dots,m$ la fonction

$$\mathbb{C}^n \ni \omega \mapsto g_r(x_o,\omega;\xi')$$

est un polynôme de degré $\leqslant k(m-r)$, où $k \in \mathbb{N}$. Choisissons

(3.5) $$R(\omega) = \sqrt{1+|\omega|^2} \; ; \quad \text{donc} \quad |\nabla R| < 1 .$$

Le choix $B[R] = R^k$ implique que Φ est admissible.

La définition (3.4) de Δ est donc :

(3.6) $$\Delta = \{(\sigma,\omega) \in \Sigma \times \mathbb{C}^n \; ; \; (k-1)\, R^{k-1}(\omega).\, \mathrm{dist}(\sigma,\alpha) < 1\} .$$

Note

Si $k = 0$ et si $k = 1$, alors $\Delta = \Sigma \times \Omega$.

4. Un énoncé complétant le précédent - Soient deux fonctions de classe C^1

$$R : \Omega \to \dot{\mathbb{R}}_+ \;, \; S : \Omega \to \dot{\mathbb{R}}_+$$

telles que

(4.1) $$\lim_{\omega \to \partial\Omega} [R(\omega) + S(\omega)] = \infty .$$

Soient deux fonctions

$$B[.,.] : \dot{\mathbb{R}}_+^2 \to \dot{\mathbb{R}}_+ \;, \quad C[.,.] : \dot{\mathbb{R}}_+^2 \to \dot{\mathbb{R}}_+$$

telles que les fonctions

(4.2) $$S \mapsto B[R,S] \quad \text{et} \quad R \mapsto C[R,S] \quad \text{croissent.}$$

Notons

$$B(\omega) = B[R(\omega), S(\omega)] \;, \quad C(\omega) = C[R(\omega), S(\omega)] .$$

Nous pouvons multiplier $B[.,.]$ et $C[.,.]$ par une même fonction de $R + S$ croissant suffisamment vite pour que

$$(4.3) \qquad \Phi = \{(\omega, \frac{\theta}{B(\omega)} \nabla R(\omega) + \frac{1-\theta}{C(\omega)} \nabla S(\omega) \in T^{\star}(\Omega) \ ; \ \omega \in \Omega \ , \ \theta \in [0,1] \ \}$$

soit admissible.

Soit $\Gamma[R,S]$ l'arc orienté unique, appartenant à $\dot{R}^2_+$, d'origine (R,S), maximal, tel que

$$(4.4) \qquad \frac{dR'}{B[R',S']} = \frac{dS'}{C[R',S']}$$

quand (R,S') décrit cet arc ; quand (R',S') tend l'extrémité de cet arc, on a donc $\lim(R'+S')=\infty$.

Définissons

$$(4.5) \qquad F[R,S] = \int_{\Gamma[R,S]} \frac{dR'}{B[R',S']} = \int_{\Gamma[R,S]} \frac{dS'}{C[R',S']} \leqslant \infty \ .$$

Cette fonction F croit avec chacun de ses arguments. Elle est continue ou identique à ∞ sur $\dot{R}^2_+$

Notons

$$F(\omega) = F\ [R(\omega)\ ,\ S(\omega)\ ]\ .$$

<u>PROPOSITION II</u>

<u>Sous les conditions précédentes, le problème</u> (1) <u>possède une unique solution holomorphe sur le domaine</u>

$$(4.6) \qquad \Delta = \{(\sigma,\omega) \in \Sigma \times \Omega \ ; \ \mathrm{dist}(\sigma,\alpha) < F(\omega)\} \ ,$$

<u>si les données</u> a <u>et</u> v <u>du problème</u> (1) <u>sont holomorphes sur ce domaine</u> Δ.

<u>Note</u>

Si $F = \infty$, on a donc $\Delta = \Sigma \times \Omega$.

<u>Exemple II</u>

<u>Soit</u> $\mathbb{P}^n$ <u>l'espace projectif complexe de dimension</u> n ; <u>soit</u> $\mathbb{P}^{n-1}$ <u>l'un de ses hyperplans ; soit</u> $\mathbb{C}^n = \mathbb{P}^n \setminus \mathbb{P}^{n-1}$ <u>l'espace vectoriel complexe de dimension</u> n ; <u>soit</u> Π <u>une hypersurface algébrique de</u> $\mathbb{P}^n$ <u>ne contenant pas</u> $\mathbb{P}^{n-1}$; <u>soit</u> p <u>son degré</u> ; <u>soit</u> P <u>un polynôme</u> $\mathbb{C}^n \to \mathbb{C}$, <u>de degré</u> p , <u>tel que</u>

$$\Pi \cap \mathbb{C}^n = \{\ \varpi \in \mathbb{C}^n \ ; \ P(\varpi) = 0\ \}\ .$$

<u>Munissons</u> $\mathbb{C}^n$ <u>d'une structure hermitienne ; soit</u>

$$(4.7) \qquad R(\varpi) = \sqrt{1 + |\varpi|^2} \quad \text{où} \quad \varpi \in \mathbb{C}^n\ .$$

Choisissons le polynôme P tel que

$$(4.8) \qquad |P(\varpi)| \leqslant R^p(\varpi) \quad , \quad \left|\frac{\partial P(\varpi)}{\partial \varpi}\right| \leqslant p\, R^{p-1}(\varpi) \, .$$

Soit Ω un revêtement d'ordre fini de $\mathbb{C}^n \setminus \Pi \cap \mathbb{C}^n$; notons ϖ la projection canonique de $\omega \in \Omega$ sur $\mathbb{C}^n$; notons

$$R(\omega) = R(\varpi) \;,\; P(\omega) = P(\varpi) \,.$$

Supposons que $g_m(x_o,\omega) = g_m(x_o)$ est indépendant de ω et que, pour $r = 0,\dots,m-1$, la fonction

$$\omega \mapsto g_r(x_o,\omega;\xi')$$

est algébrique et holomorphe sur Ω ; il existe donc k et $\ell \in \mathbb{R}_+$ tels

$$(4.9) \qquad \Phi = \{(\omega, \theta \frac{|P^\ell(\omega)|}{R^k(\omega)} \nabla R(\omega) - (1-\theta) \frac{|P^{\ell-1}(\omega)|\; \overline{P(\omega)}}{pR^{k+p-1}(\omega)} \frac{\partial P}{\partial \omega} \;;\; \omega \in \Omega,\ \theta \in [0,1]\}$$

est admissible.

Définissons, pour $Q \in \mathbb{R}_+$,

$$(4.10) \qquad I\,[Q] = \frac{1}{p} \int_o^Q (Q-t)^\ell (1+t)^{-(k+p-1)/p}\, dt \;;$$

vu (4.8), la fonction

$$(4.11) \qquad \mathbb{C}^n \ni \varpi \to I\left[|P(\varpi)| / R^p(\varpi)\right] = J(\varpi) \in \mathbb{R}_+$$

se prolonge en une fonction bornée et continue

$$\mathbb{P}^n \ni \varpi \to J(\varpi) \in \mathbb{R}_+$$

telle que

$$\Pi = \{\, \varpi \in \mathbb{P}^n \;;\; J(\varpi) = 0 \,\} \,.$$

Soit

$$(4.12) \qquad F(\varpi) = J(\varpi)\, R^{\ell p+1-k}(\varpi) \quad ; \; F(\omega) = F(\varpi) \quad \text{pour } \omega \in \Omega \quad .$$

Le problème (1) possède une solution holomorphe sur le domaine

$$(4.13) \qquad \Delta = \{(\sigma,\omega) \in \Sigma \times \Omega \quad ; \; \mathrm{dist}(\sigma,\alpha) < F(\omega)\} \,,$$

si les données a et v sont holomorphes sur ce domaine Δ.

Cas particulier

Supposons que, pour $r = 0,\dots,m$, la fonction

$$\varpi \mapsto g_r(x_o,\varpi\,;\xi')$$

est un polynome de degré $\leqslant k(m-r)$. Alors on peut choisir $\ell = 0$ et choisir pour k le nombre précédent ; la définition (4.12) de F devient :

$$(4.14) \qquad F(\omega) = \frac{1}{1-k}\left\{\left[\left|P(\omega)\right| + R^{p}(\omega)\right]^{\frac{1-k}{p}} - R^{1-k}(\omega)\right\} \quad \text{si} \quad k \neq 1 ,$$

$$= \frac{1}{p}\log\frac{|P(\omega)| + R^{p}(\omega)}{R^{p}(\omega)} \qquad \text{si} \quad k = 1 .$$

5. Emploi d'une métrique sur Ω .

Munissons Ω d'une métrique, définie par exemple par un $ds^{2}(\omega)$ riemannien. Elle définit la norme $|\xi'|$ de tout $\xi' \in T^{\star}_{\omega}(\Omega)$. Il existe une fonction

$$\mu : \Omega \to \dot{\mathbb{R}}_{+}$$

telle que

$$(5.1) \qquad \Phi = \{(\omega;\xi') \in T^{\star}(\Omega) \ ; \ |\xi'| = \mu(\omega)\}$$

soit admissible.

Notons

$$(5.2) \qquad dS(\omega) = \mu(\omega) \quad ds(\omega)$$

et nommons $\mathrm{Dist}(\mu,\partial\Omega)$ la distance, pour la métrique $dS^{2}(\omega)$, de ω au point à l'infini $\partial\Omega$ de Ω ; c'est-à-dire :

$$(5.3) \qquad \mathrm{Dist.}(\omega,\partial\Omega) = \inf \int_{\omega}^{\partial\Omega} \mu(\omega)\, ds(\omega) \leqslant \infty .$$

Proposition III - Sous les conditions précédentes, le problème (1) possède une unique solution holomorphe sur le domaine

$$(5.4) \qquad \Delta = \{(\sigma,\omega) \in \Sigma \times \Omega \ ; \ \mathrm{dist.}(\sigma,\alpha) < \mathrm{Dist}(\omega,\partial\Omega)\},$$

si les données a et v sont holomorphes sur ce domaine Δ .

Note

Si $\mathrm{Dist.}(\omega,\partial\Omega) = \infty$ en un point ω de Ω , donc en tout point ω de Ω , c'est-à-dire si Ω est un espace complet pour la métrique $\mu^{2}(\omega)\, ds^{2}(\mu)$, alors $\Delta = \Sigma \times \Omega$.

Note

[H.L.T.2] donnera des conséquences de cette proposition apparentées aux propositions I et II .

BIBLIOGRAPHIE

[H.T.] Y. HAMADA et T. TAKEUCHI, C.R. Acad. Sc. Paris, 295, Série I, 1982, p. 329-332.

[H.L.T.,1] Y. HAMADA, J. LERAY et T. TAKEUCHI, C.R. Acad. Sc. Paris, 296 série I, 1983, p. 435-437.

[H.L.T.,2] Y. HAMADA, J. LERAY et T. TAKEUCHI, à paraître au J. Math. Pures et Appliquées.

[L.1] J. LERAY, Bull. Soc. Math. Fr. 85, 1957, p. 389-429.

[L.2] J. LERAY, Congrès dédié à E. MARTINELLI (Rome, 1983), Riv. Mat. Univ. Parma (4) 10.

Jean LERAY
Professeur
Collège de France
Place Berthelot
75231 - PARIS CEDEX 05

P L LIONS

Equations de Hamilton–Jacobi et solutions de viscosité

Introduction

Dans de nombreux domaines des Mathématiques et de la Physique (Calcul des variations, contrôle optimal, jeux différentiels, grandes déviations, optique géométrique, approximation semi-classique ...) interviennent les équations de Hamilton-Jacobi (H J en abrégé) i.e. :

(1) $$F(x,u(x), \nabla u(x)) = 0 \quad \text{dans } \mathcal{O}$$

où $F \in C(\bar{\mathcal{O}} \times \mathbb{R} \times \mathbb{R}^m)$ est donnée (F est l'hamiltonien), $\mathcal{O}$ est un ouvert de $\mathbb{R}^m$, $m \geqslant 1$ et u - la fonction inconnue - est scalaire.

Cette équation complètement non-linéaire du premier ordre, n'admet pas en général de solutions régulières : plus précisément si on cherche une solution globale de (1) avec des conditions aux limites données sur $\partial\mathcal{O}$, il n'existe pas en général de solutions de classe C^1. Ainsi le problème

(2) $$|\nabla u| = 1 \quad \text{dans } \mathcal{O} \quad , \quad u = 0 \quad \text{sur } \partial\mathcal{O}$$

avec $\mathcal{O}$ borné par exemple) n'admet aucune solution de classe C^1.

Il est alors naturel de chercher des solutions u de (1) seulement localement lipschitziennes ; l'équation étant alors satisfaite p.p. . Cette approche est celle suivie par A. DOUGLIS [19] ; S.N. KRUZKOV [29], [30], [31] ; W.H. FLEMING [23], [24] ; A. FRIEDMAN [26] ... L'existence de solutions localement Lipschitziennes est obtenue sous des hypothèses générales sur F mais l'unicité et la stabilité ne sont pas vérifiées : ainsi dans l'exemple (2), pour toute partition de $\bar{\mathcal{O}}$ $(F_i)_{1 \leqslant i \leqslant k}$ on pose $\mathcal{O}_i = (\mathbb{R}^m - F_i) \cap \bar{\mathcal{O}}$ et on vérifie que

$$u(x) = \varepsilon_i \operatorname{dist}(x,\mathcal{O}_i) \quad , \quad \forall x \in F_i \ , \ \forall\, 1 \leqslant i \leqslant k \ , \ \varepsilon_i = \pm 1 \ ;$$

vérifie (2) p.p. Et par des choix convenables de $k, (F_i)$: on obtient une suite de solutions $(u_n)_n$ de (2) convergeant uniformément vers 0 qui n'est bien sûr pas solution de (2) .

Récemment M.G. CRANDALL et l'auteur ont introduit une notion de solutions faibles de (1) qui permet de résoudre complètement les équations de H J : dans [14] , [15] les principales propriétés de ces solutions - appelées solutions de viscosité - et des résultats d'unicité et de comparaison sont démontrés (voir aussi M.G. CRANDALL, L.C. EVANS et P.L. LIONS [13] , P.L. LIONS [32]). Nous rappelons dans la section 1 la définition de ces solutions de viscosité ainsi que leurs principales propriétés. La section 2 est consacrée à l'existence et l'unicité de solutions de viscosité pour des problèmes modèles. Ces rappels sont brefs et nous renvoyons le lecteur à la bibliographie pour plus de références. Ainsi que nous essaierons de le montrer l'outil des solutions de viscosité est d'une utilisation simple et a permis de résoudre de nombreux problèmes concernant les diverses applications des équations de Hamilton-Jacobi. Sans souci d'exhaustivité signalons les sujets suivants qui ont pu être traités grâce aux solutions de viscosité : i) résultats d'existence : P. LIONS [32] , [33] , P.E. SOUGANIDIS [43] , G. BARLES [3],[4] x...,ii) approximation numérique : M.G. CRANDALL et P.L. LIONS [16] , P.E. SOUGANIDIS [43] , I. CAPUZZO-DOLCETTA [11]; iii) existence et unicité dans des cas non-bornés : H. ISHII [27] , M.G. CRANDALL et P.L. LIONS [17] ; iv) étude des semi-groupes : P.L. LIONS et M. NISIO [39] , P.L. LIONS [34] ; v) propriétés qualitatives : BARDI et L.C. EVANS [2], BARDI [1] , P.L. LIONS et J.C. ROCHET [41] ; vi) liens avec le contrôle optimal et les jeux différentiels : P.L. LIONS [32] , P.E. SOUGANIDIS [44] , L.C. EVANS et P.E. SOUGANIDIS [22], E.N. BARRON, L.C. EVANS et R. JENSEN [7] , L.C. EVANS et H. ISHII [20] , P.L. LIONS et P.E. SOUGANIDIS [42] , G. BARLES [5] , [6] , I.CAPUZZO-DOLCETTA et L.C. EVANS [12] , vii) théorie des grandes déviations : L.C. EVANS et H. ISHII [20] , W.H. FLEMING et P.E. SOUGANIDIS [25] ; viii) extension au 2ème ordre et contrôle optimal stochastique : P.L. LIONS [35] , [36] , [37] ...

Les sections [3] et [4] sont consacrées à deux autres applications des solutions de viscosité : il s'agit d'une part de problèmes asymptotiques de type homogénéisation qui sont tirés de P.L. LIONS, G. PAPANICOLAOU et S.R.S. VARADHAN [40] et d'autre part d'un problème de perturbations singulières adapté de R. JENSEN et P.L. LIONS [28] .

1 - Solutions de viscosité : définitions et propriétés

On rappelle que si $\varphi \in C(\mathcal{O})$ et si $X \in \mathcal{O}$, le surdifférentiel de φ au point x - noté $D^+\varphi(x)$ - est l'ensemble convexe fermé éventuellement vide défini par :

$$D^+\varphi(x) = \{\xi \in \mathbb{R}^m \ / \ \limsup_{y \to x, y \in \mathcal{O}} \{ \varphi(y) - \varphi(x) - (\xi, y-x)\} \, |y-x|^{-1} \leqslant 0 \}$$

ou bien $\quad = \{\xi \in \mathbb{R}^m \ / \ \exists \psi \in C^1(\mathcal{O}) \ , \ \psi(x) = \varphi(x) \ , \ \nabla\psi(x) = \xi \ ,$

$$\psi(y) > \varphi(y) \quad \text{si} \quad y \in \mathcal{O} \ , \ y \neq x \} .$$

De même définit-on le sous-différentiel de φ au point x - on note $D^-\varphi(x)$ - en remplaçant lim sup par lim inf et " $\leqslant$ " par " $\geqslant$ " de sorte que :

$$D^-\varphi(x) = - D^+(-\varphi)(x) .$$

Définition

Soit $u \in C(\mathcal{O})$, on dit que u est

i) sous-solution de viscosité de (1) si

(3) $\quad F(x, u(x), \xi) \leqslant 0 \ , \ \forall \ x \in \mathcal{O} \ , \ \forall \ \xi \in D^+u(x)$

ii) sur-solution de viscosité de (1) si

(4) $\quad F(x,u(x), \xi) \geqslant 0 \ , \ \forall \ x \in \mathcal{O} \ , \ \forall \ \xi \in D^-u(x)$

iii) solution de viscosité de (1) si u est à la fois sous-solution et sur-solution de viscosité de (1).

Remarques

Toute solution de classe C^1 de (1) est solution de viscosité et toute solution de viscosité de (1) vérifie l'équation (1) en tout point de differentiabilité.

Exemple

Dans l'exemple donné par (2) , $u(x) = \text{dist}(x, \partial\mathcal{O})$ est solution de viscosité de (2) et nous verrons plus loin que c'est la seule.

Définition équivalente

$u \in C(\mathcal{O})$ est sous-solution (resp. sur-solution, resp-solution) de viscosité de (1) si et seulement si on a pour tout $\psi \in C^1(\mathcal{O})$

(5) en tout point x_o de maximum local de $u - \psi$: $F(x_o,u(x_o), D\psi(x_o)) \leqslant 0$ (resp.

(6) en tout point x_o de minimum local de $u-\psi$: $F(x_o,u(x_o), D\psi(x_o)) \geqslant 0$; resp. (5) et (6).

Remarques

On obtient encore des définitions équivalentes en remplaçant C^1 par C^2, C^∞ ...; ou en remplaçant local par global, local strict, global strict...

La terminologie utilisée provient de la remarque (fondamentale) suivante : soit $u_\varepsilon \in C^2(\mathcal{O})$ solution de

(7) $$-\varepsilon \Delta u_\varepsilon + F_\varepsilon(x,u_\varepsilon,Du_\varepsilon) = 0 \quad \text{dans } \mathcal{O}$$

où $\varepsilon > 0$, $F_\varepsilon(x,t,p) \in C(\mathcal{O}\times\mathbb{R}\times\mathbb{R}^m)$ converge vers F uniformément sur tout compact de $\mathcal{O}\times\mathbb{R}\times\mathbb{R}^m$ quand $\varepsilon \to 0$. Supposons que u_ε converge uniformément vers une fonction u sur tout compact de $\mathcal{O}$ quand $\varepsilon \to 0$ (ou une sous-suite...) - ceci sera le cas si on a obtenu une estimation a priori sur $\|\nabla u_\varepsilon\|_{L^\infty_{loc}}$ - .

Alors u est solution de viscosité de (1) ; en effet vérifions par exemple que u est sur-solution de viscosité i.e. que (6) a lieu pour $\psi \in C^2$ et pour x_o point de minimum local strict de $u-\psi$: or pour ε petit $u_\varepsilon - \psi$ admet un point de minimum local x_ε qui converge vers x_o. De plus $\nabla u_\varepsilon(x_\varepsilon) = \nabla\psi(x_\varepsilon)$, $\Delta u_\varepsilon(x_\varepsilon) \geqslant \Delta\psi(x_\varepsilon)$; d'où

$$F_\varepsilon(x_\varepsilon,u_\varepsilon(x_\varepsilon), \nabla\psi(x_\varepsilon)) = \varepsilon\Delta u_\varepsilon(x_\varepsilon) \geqslant \varepsilon\Delta\psi(x_\varepsilon)$$

et on conclut en faisant tendre ε vers 0.

Un raisonnement du même type montre que si $u_n \in C(\mathcal{O})$ est sous-solution (resp. sur-solution, resp. solution) de viscosité de

$$F_n(x,u_n,Du_n) = 0 \quad \text{dans } \mathcal{O}$$

et si u_n, F_n convergent uniformément sur tout compact vers u,F alors u est sous-solution (resp. sur-solution, resp. solution) de viscosité de (1).

2 - Résultats d'existence, d'unicité et de comparaison

Afin de simplifier la présentation, nous ne considérerons ici que les cas modèles suivants :

(8) $$H(x,u,Du) = 0 \quad \text{dans } \mathbb{R}^N ;$$

(9) $\frac{\partial u}{\partial t} + H(x,u,Du) = 0$ dans $\mathbb{R}^N \times]0,T[= Q$;

bien sûr (9) est un cas particulier de (1) en remplaçant N par N+1, $\mathcal{O}$ par Q, x par (x,t) et F par $p_{N+1} + H(x, \lambda, p')$ (où $p = (p', p_{N+1}) \in \mathbb{R}^N \times \mathbb{R}$).

Commençons par (8) : nous utiliserons les hypothèses suivantes

(10) $H(x,t,p) \in BUC(\mathbb{R}^N \times [-R,+R] \times \overline{B}_R)), \ \forall \ R < \infty$;

(11) $\exists \lambda > 0, \ \forall (x,p) \in \mathbb{R}^N \times \mathbb{R}^N, \ \forall t,s \in \mathbb{R}$

$$(H(x,t,p) - H(x,s,p))(t-s) \geq \lambda(t-s)^2$$

(12) $|H(x,t,p) - H(y,t,p)| \leq C_R^1 |x-y| \ |p| + C_R^2 |x-y|$

$$\forall \ x,y,p \in \mathbb{R}^N, \ \forall t \in [-R,+R], \ \forall R < \infty \ ;$$

(13) $|H(x,t,p) - H(y,t,p)| \leq \omega_R(|x-y|(1+|p|)$ avec $\omega_R(t) \to 0$ si $t \to 0_+$; $\forall x,y,p \in \mathbb{R}^N$, $\forall t \in [-R,+R]$, $\forall R < \infty$.

(Remarquer que (12) entraîne (13)) ;

(14) $H(x,t,p) \to +\infty$ si $|p| \to +\infty$, uniformément pour $x \in \mathbb{R}^N$, t borné.

Théorème 1

i) Soient $u,v \in BUC(\mathbb{R}^N)$ resp. sous et sur-solution de viscosité de (8),

(8') $H(x,v,Dv) = f$ dans $\mathbb{R}^N$ où $f \in C_b(\mathbb{R}^N)$. On suppose (10), (11) et soit (13), soit (14) alors on a

(15) $$\sup_{\mathbb{R}^N} (u-v)^+ \leq \frac{1}{\lambda} \sup_{\mathbb{R}^N} f^- .$$

ii) On suppose (10), (11 et soit (13) soit (14). Alors il existe une unique solution u de viscosité de (8) dans $BUC(\mathbb{R}^N)$. Si (14) a lieu, $u \in W^{1,\infty}(\mathbb{R}^N)$ tandis que si (12) a lieu et si on note $\overline{R} = \|u\|_\infty$, $C_o = C^1_{\overline{R}}$ alors $u \in C^{o,\gamma}(\mathbb{R}^N)$ avec $\gamma = \lambda/C_o$ si $\lambda < C_o$, $\gamma \in]0,1[$ si $\lambda = C_o$, $\gamma = 1$ si $\lambda > C_o$.

Dans le cas de (9), on remplace (11) par

(11') $\exists\ \lambda \in \mathbb{R}, \forall\ x,p \in \mathbb{R}^N\ , \forall\ t,s \in \mathbb{R}$

$$(H(x,t,p) - H(x,s,p))(t-s) \geqslant \lambda(t-s)^2 .$$

Théorème 2

i) Soient $u,v \in BUC(\overline{Q})$ resp. sous et sur-solution de viscosité de (9),

(9') $$\frac{\partial v}{\partial t} + H(x,v,Dv) = f \quad \text{dans} \quad Q$$

où $f \in C_b(\overline{Q})$. On suppose (10), (11') et soit (13), soit (14) ; alors

(16) $$\sup_{x\in\mathbb{R}^N} (u(x,t)-v(x,t))^+ \leqslant \sup_{x\in\mathbb{R}^N} (u(x,0)-v(x,0))^+ \ e^{-\lambda t} + \\ + \int_o^t \sup_{x\in\mathbb{R}^N} f^-(x,s)\ e^{+\lambda s}\ ds\ e^{-\lambda t} .$$

ii) On suppose (10),(11') et soit (13), soit (14). Alors il existe une unique solution de viscosité u de (9) dans $BUC(\overline{Q})$ vérifiant la condition initiale

(17) $$u\big|_{t=o} = u_o$$

où $u_o \in BUC(\mathbb{R}^N)$. De plus si (12) a lieu et $u_o \in C^{o,\gamma}(\mathbb{R}^N)$ avec $0 < \gamma \leqslant 1$ alors $u \in C^{o,\gamma}(\overline{Q})$; et si (14) a lieu et $u_o \in W^{1,\infty}(\mathbb{R}^N)$ alors $u \in W^{1,\infty}(Q)$.

Remarques

i) Les résultats d'unicité et de comparaison sont dûs à M. G. GRANDALL et P.L. LIONS [15]. Les résultats d'existence dans le cas où soit (12), soit (14) ont lieu ainsi que les résultats de régularité sont dûs à P.L. LIONS [32], [33], P.E. SOUGANIDIS [43]; le cas général (hypothèse (13)) étant dû à G. BARLES [3].

ii) Dans [15] sont données de nombreuses variantes, extensions et adaptations de ces résultats. Signalons ainsi que si $\mathcal{O}$ est un ouvert borné de $\mathbb{R}^N$, si (10), (11), (13) ont lieu et si $u,v \in C(\overline{\mathcal{O}})$ sont des solutions de viscosité de

(18) $$H(x,u,Du) = 0 \quad \text{dans } \mathcal{O}$$

alors $\max_{\overline{\mathcal{O}}} (u-v)^+ = \max_{\partial\mathcal{O}} (u-v)^+$.

Dans le cas où $H(x,t,p)$ ne dépend pas de t : $H = H(x,p)$, comme c'est le cas pour le problème (2), l'unicité n'a pas lieu en général sauf si on a (10), (13) ou (14) et

(19) $H(x,p)$ est convexe en p , $H(x,0) < 0$ sur $\bar{\mathcal{O}}$.

Noter que (19) a lieu dans l'exemple (2) puisque $H(x,p) = |p| - 1$.

3 - Un problème asymptotique

Les résultats de cette section sont tirés de P.L. LIONS, G. PAPANICOLAOU et S.R.S. VARADHAN [40] : on s'intéresse au comportement quand ε tend vers 0 de la solution de viscosité $u^\varepsilon \in BUC(\mathbb{R}^N \times [0,T])$ $(\forall T < \infty)$ de

(20) $$\frac{\partial u^\varepsilon}{\partial t} + H\left(\frac{x}{\varepsilon}, Du^\varepsilon\right) = 0 \quad \text{dans} \quad \mathbb{R}^N \times]0,\infty[$$

(21) $$u^\varepsilon\big|_{t=0} = u_o \quad \text{sur} \quad \mathbb{R}^N$$

où $u_o \in BUC(\mathbb{R}^N)$, $H(x,p) \in C(\mathbb{R}^N \times \mathbb{R}^N)$ est périodique de période 1 par rapport à x_i $(1 \leqslant i \leqslant N)$ et vérifie

(22) $$H(x,p) \to +\infty \quad \text{si} \quad |p| \to \infty \text{ , uniformément pour } x \in \mathbb{R}^N .$$

Un tel problème intervient notamment en optique (propagation des rayons lumineux dans des milieux non homogènes).

Un développement asymptotique formel fait apparaître que l'hamiltonien "moyenné" (ou "effectif") doit être déterminé par le problème ergodique suivant : trouver (λ,v) vérifiant :

(23) $$\begin{cases} \lambda \in \mathbb{R} \text{ , } v \in C(\mathbb{R}^N), v \text{ est périodique, } v \text{ est solution de viscosité} \\ \quad \text{de } H(x,Dv+p) = \lambda \text{ dans } \mathbb{R}^N \end{cases}$$

où $p \in \mathbb{R}^N$ est fixé. En posant $\lambda = \bar{H}(p)$, $\bar{H}$ doit être l'hamiltonien effectif. Il nous faut donc résoudre (23) ; pour cela on considère u_ε la solution de viscosité dans $BUC(\mathbb{R}^N)$ de :

(24) $$H(x,Du_\varepsilon+p) + \varepsilon u_\varepsilon = 0 \quad \text{dans} \quad \mathbb{R}^N .$$

D'après le théorème 1 , u_ε est nécessairement périodique et

(25) $$\| \varepsilon u_\varepsilon \|_\infty \leqslant \| H(x,p) \|_\infty$$

(on utilise ici le fait que 0 est solution de viscosité de

$$H(x,Dw+p) + \varepsilon w = H(x,p))\ ;$$

d'où d'après (22) en utilisant l'équation (et les propriétés des solutions de viscosité) : $\|Du_\varepsilon\|_\infty \leqslant C$ (ind. de ε). Cette borne combinée avec (25) implique que - quitte à extraire une sous-suite $-\varepsilon u_\varepsilon, u_\varepsilon - u_\varepsilon(0)$ convergent uniformément respectivement vers $-\lambda \in \mathbb{R}$, $v \in W^{1,\infty}(\mathbb{R}^N)$ périodique solution de viscosité de

$$(23')\qquad H(x,Dv + p) = \lambda \text{ dans } \mathbb{R}^N .$$

De plus λ est l'unique réel tel qu'il existe v périodique, $v \in C(\mathbb{R}^N)$, solution de viscosité de (23') : en effet si $\lambda > \mu$ sont deux réels tels qu'il existe de telles fonctions v_1, v_2 on a pour ε petit

$$\lambda + \varepsilon v_1 > \mu + \varepsilon v_2 \quad \text{sur } \mathbb{R}^N$$

et en remarquant que v_1, v_2 sont solutions de viscosité de

$$H(x,Dv_1 + p) + \varepsilon v_1 = \lambda + \varepsilon v_1 \quad \text{dans } \mathbb{R}^N$$

$$H(x,Dv_2 + p) + \varepsilon v_2 = \mu + \varepsilon v_2 \quad \text{dans } \mathbb{R}^N$$

on déduit du théorème 1 que : $v_1 \geqslant v_2$ sur $\mathbb{R}^N$. Et ceci est absurde puisque v_1, v_2 sont définis à une constante près ! Nous avons donc démontré la :

Proposition 3

Avec les notations précédentes, il existe, pour tout $p \in \mathbb{R}^N$, un unique réel $\lambda = H(p)$ tel qu'il existe v périodique continue solution de viscosité de (23'). De plus on a :

$$\underset{x\in\mathbb{R}^N}{\operatorname{Min}} H(x,p) \leqslant \lambda \leqslant \underset{x\in\mathbb{R}^N}{\operatorname{Max}} H(x,p)$$

et si H est convexe par rapport à p , $\lambda = H(p)$ est également convexe par rapport à p .

Remarques

Quelques propriétés qualitatives complémentaires sont données dans [40]. En général v n'est pas unique (même à une constante près).

Revenons maintenant au problème asymptotique (20) - (21) :

Théorème 4

La solution de viscosité u^{ε} de (20)-(21) converge uniformément sur $\mathbb{R}^N \times [0,T]$ $(\forall T < \infty)$ vers la solution de viscosité u dans $BUC(\mathbb{R}^N \times [0,T])$ $(\forall T < \infty)$ de

$$(26) \qquad \frac{\partial u}{\partial t} + H(Du) = 0 \quad \text{dans} \quad \mathbb{R}^N \times]0,\infty[$$

vérifiant la condition initiale (21) ; où H est déterminé ci-dessus.

La démonstration suit le schéma suivant : en utilisant des arguments convenables de densité et de propagation à vitesse finie du support, on vérifie qu'il existe $\varepsilon_n \underset{n}{\longrightarrow} 0$ (suite extraite d'une suite arbitraire) tel que le semi-groupe S_{ε_n} correspondant vérifie : $S_{\varepsilon_n}(t)\,u_o$ converge uniformément sur tout compact de $\mathbb{R}^N \times [0,\infty[$ vers $S(t)u_o$ ou $S(t)$ est un semi-groupe continu sur $BUC(\mathbb{R}^N)$, de contractions, respectant l'ordre et possédant la propriété de propagation à vitesse finie du support.

D'autre part si $u_o = \alpha + p.x$ où $\alpha \in \mathbb{R}$, $p \in \mathbb{R}^N$; on considère

$$u_o^{\varepsilon} = \alpha + p.x + \varepsilon v\left(\frac{x}{\varepsilon}\right) \text{ où } v \text{ est solution de (23').}$$

Et on remarque que

$$(S_{\varepsilon}(t)\,u_o^{\varepsilon})\,(x) = \varepsilon + p.x - t\,H(p) + \varepsilon v\left(\frac{x}{\varepsilon}\right)$$

$$\| S_{\varepsilon}(t)\,u_o - S_{\varepsilon}(t)\,u_o^{\varepsilon} \|_{\infty} \leqslant \varepsilon \| v \|_{\infty} .$$

Et ceci entraîne : $S(t)\,u_o = \alpha + p.x - t\,H(p)$.

On déduit alors aisément des propriétés de propagation de support et de l'action de $S(t)$ sur les fonctions affines que :

$$\frac{1}{t}\{S(t)\varphi - \varphi\}\,(x) \underset{t \to o_+}{\longrightarrow} -H(D\varphi(x)) , \quad \forall x \in \mathbb{R}^N$$

pour tout $\varphi \in C_b^1(\mathbb{R}^N)$, $D\varphi \in BUC(\mathbb{R}^N)$ (et la convergence est uniforme par rapport à φ si φ est bornée dans $C_b^2(\mathbb{R}^N)$ par exemple).

On conclut alors en appliquant le résultat de P.L. LIONS et M. NISIO [39].

Remarque

Les hypothèses du type (22) interviennent souvent dans la théorie des équations de Hamilton-Jacobi et ce car pour toute sous-solution de viscosité u de

$$H(x,Du) \leqslant C \quad \text{dans} \quad \mathbb{R}^N \ , \ u \in C(\mathbb{R}^N)$$

vérifie automatiquement : $| u(x) - u(y)| \leqslant K(C) \ |x-y| \ , \forall x,y$.

En fait on peut généraliser considérablement (22) ; l'exemple suivant indique clairement une généralisation possible :

$$H(x_1,x_2,p_1,p_2) = \text{Max} \ (x_1(1-x_1)|p_2|, \ |p_1|)$$

si $0 \leqslant x_1,x_2 \leqslant 1$, $p_1,p_2 \in \mathbb{R}$. Alors pour tout $u \in C(\overline{\pi})$ vérifiant :

$$H(x,Du) \leqslant M \quad \text{(au sens de viscosité)}$$

on a : $|u(x) - u(y)| \leqslant C_o \ M \ |x-y|^{1/2} \ , \ \forall \ x,y \in \overline{\pi} = \{0 \leqslant x_1,x_2 \leqslant 1\}$

où C_o ne dépend ni de u , ni de M . En fait l'exemple précédent est un cas typique d'hypoellipticité (et $\alpha = 1/2$ correspond au nombre de crochets de Lie nécessaire pour "remplir le rang" de l'algèbre de Lie). De tels arguments sont développés dans BENAROUS et P.L. LIONS [8].

4 - Un problème de perturbations singulières

On s'intéresse maintenant au comportement, quand ε tend vers 0 , de la solution u^ε de

$$(27) \quad \begin{cases} \dfrac{\partial u^\varepsilon}{\partial t} + H(x,y, D_x u^\varepsilon, \dfrac{1}{\varepsilon} D_y u^\varepsilon) = 0 & \text{dans } \mathbb{R}^n \times \mathbb{R}^m \times \ ]0,\infty[\\ u^\varepsilon|_{t=o} = u_o(x,y) & \text{sur } \mathbb{R}^n \times \mathbb{R}^m \end{cases}$$

où $H(x,y,p,q) \in BUC(\mathbb{R}^n \times \mathbb{R}^m \times \overline{B}_R \times \overline{B}_R) \quad (\forall R < \infty)$.

La motivation de ce problème est l'étude de grands systèmes où des sous-systèmes évoluent à des vitesses différentes et leur contrôle optimal. Dans ce contexte H est donné par :

$$(28) \qquad H(x,y,p,q) = \sup_{\alpha \in \mathcal{A}} \{ -b_\alpha(x,y).p - \beta_\alpha(x,y).q - f_\alpha(x,y) \} .$$

ou $\mathcal{A}$ est un ensemble donné (ensemble des valeurs du contrôle), b_α ,β_α sont bornés dans $W^{1,\infty}(\mathbb{R}^n \times \mathbb{R}^m)$, f_α est borné dans $C_b(\mathbb{R}^n \times \mathbb{R}^m)$ et f est uniformément continu sur $\mathbb{R}^n \times \mathbb{R}^m$ uniformément en α . Pour plus de détails nous renvoyons le lecteur à A. BENSOUSSAN et G. BLANKENSHIP [10] , A. BENSOUSSAN [9] , P.L. LIONS [32] ...

Le problème général (27) - même pour l'hamiltonien (28) - reste largement ouvert : nous présentons ici sa résolution dans un cas particulier adapté de R. JENSEN et P.L. LIONS [28] . Nous supposerons :

(29) $|\tilde{H}(x_1,y,p,q) - \tilde{H}(x_2,y,p,q)| \leqslant C\,|x_1 - x_2|\,(1+ |p|)\ ,\ \forall x_1,x_2,y,p,q$

où $\tilde{H}(x,y,p,q) = H\,(x,y,p,q) - H(x,y,0,0)$;

(30) $H \to +\infty$ si $|q| \to \infty$, uniformément pour $x \in \mathbb{R}^n$, $y \in \mathbb{R}^m$, p borné;

(31) $H(x,y,p,q) \geqslant H(x,y,p,o)$, $\forall\ x,y,p,q$.

Sous les hypothèses (29), (30) , il existe une unique solution $u^\varepsilon \in BUC(\mathbb{R}^N \times [0,T])$ ($\forall T < \infty$) solution de viscosité de (27) si $u_o \in BUC(\mathbb{R}^N)$ ($N = n+m$).

De plus si $u_o \in W^{1,\infty}(\mathbb{R}^N)$, $u^\varepsilon \in W^{1,\infty}(\mathbb{R}^N \times]0,T[)$ ($\forall\ T < \infty$).

Théorème 5

Sous les hypothèses (29), (30), (31) et si $u_o = u_o(x) \in BUC(\mathbb{R}^N)$ la solution u^ε de (27) converge uniformément sur tout compact de $\mathbb{R}^N \times [0,\infty[$ vers la solution de viscosité $u \in BUC(\mathbb{R}^n \times [0,T])$ ($\forall T<\infty$) de :

(32) $$\begin{cases} \dfrac{\partial u}{\partial t} + \sup\limits_{y \in \mathbb{R}^m} H(x,y,D_x u,o) = 0 & \text{dans } \mathbb{R}^n \times]0,\infty[\\ u|_{t=o} = u_o(x) & \text{sur } \mathbb{R}^n . \end{cases}$$

Remarques

i) Si u_o dépend de y , la convergence est uniforme sur tout compact de $\mathbb{R}^N \times]0,\infty[$ et la condition initiale est : $u|_{t=o} = \inf\limits_{y \in \mathbb{R}^m} u_o(x,y)$

ii) Le problème de la vitesse de convergence et des correcteurs est ouvert.

iii) Si H est donné par (28), (29) est vérifié si $\beta_\alpha = \beta_\alpha(y)$, (30) est vérifié si $\overline{co}\,\{\beta_\alpha(x,y)/\alpha \in \mathcal{A}\} \supset \overline{B}_\delta$ où $\delta > 0$ est indépendant de $x \in \mathbb{R}^n$, $y \in \mathbb{R}^m$. L'hypothèse (31) est la plus restrictive : l'exemple le plus simple est le suivant

$\mathcal{A} = \mathcal{A}_1 \times \mathcal{A}_2$, $\alpha = (\alpha_1, \alpha_2)$, $b_\alpha(x,y) = b_{\alpha_1}(x,y)$, $f_\alpha(x,y) = f_{\alpha_1}(x,y)$

$$0 \in \overline{co}\{\beta_{(\alpha_1,\alpha_2)}(x,y) \;/\; \alpha_2 \in \mathcal{A}_2\},\ \forall\ x,y,\alpha_1 .$$

<u>Esquissons la démonstration du Théorème 5</u> : Il suffit de considérer le cas ou $u_0 \in W^{1,\infty}(\mathbb{R}^n)$.

Dans ce cas on vérifie aisément :

$$\left|\frac{\partial u^\varepsilon}{\partial t}\right| \leq C \ ,\ |D_x u^\varepsilon| \leq C$$

où C est indépendante de ε. En utilisant l'équation et (30) on en déduit :

$$|D_y u^\varepsilon| \leq C\,\varepsilon (\text{avec } C \text{ indép. de } \varepsilon).$$

D'après (31), si u^ε converge uniformément sur tout compact de $\mathbb{R}^N \times [0,\infty[$ vers $\tilde{u} \in BUC(\mathbb{R}^N \times [0,T])$ $(\forall\ T < \infty)$, $\tilde{u}$ est indépendante de y et $\tilde{u}$ est une sous-solution de viscosité de (32) donc $\tilde{u} \leq u$ sur $\mathbb{R}^n \times [0,\infty[$. D'autre part u-solution de (32) - est, toujours d'après (31), une sous-solution de viscosité de (27) et donc $u \leq u^\varepsilon$ sur $\mathbb{R}^N \times [0,\infty[$. Et on conclut.

BIBLIOGRAPHIE

[1] M. BARDI, Preprint.

[2] M. BARDI et L.C. EVANS, On Hopf's formulas for solutions of Hamilton-Jacobi equations, Preprint.

[3] G. BARLES , Existence results for first order, Hamilton-Jacobi equations. A paraître aux Ann. I.H.P. Anal. non lin. voir aussi Thèse de 3ème cycle, Paris Dauphine, 1983.

[4] G. BARLES, Some remarks on existence results for first order Hamilton-Jacobi equations. Ann. I.H.P. Anal. non lin.

[5] G. BARLES, Quasivariational inequalities and first-order Hamilton-Jacobi equations. Thèse de 3ème cycle, Paris-Dauphine, 1983.

[6] G. BARLES, Detaministic impulsive control problems. A paraître au SIAM J. Control Optim. ; voir aussi Thèse de 3ème cycle, Paris-Dauphine, 1983.

[7] E.N. BARRON, L.C. EVANS et R. JENSEN, Viscosity solutions of Issac's equations and differential games with Lipschitz controls. A paraître dans J. Diff. Eq.

[8) BENAROUS, et P.L. LIONS, Travail en préparation.

[9] A. BENSOUSSAN, Livre à paraître sur les perturbations singulières en contrôle optimal.

[10] A. BENSOUSSAN et G. BLANKENSHIP, en préparation.

[11] I. CAPUZZO-DOLCETTA, On a discrete approximation of the Hamilton-Jacobi equation of dynamic programming. Appl. Math. Optim., 10 (1983), p. 367-377.

[12] I. CAPUZZO-DOLCETTA et L.C. EVANS, Optimal suitching for ordinary differential equations. SIAM J. Control Optim., 22 (1984) p.143-161

[13] M. CRANDALL, L.C. EVANS et P.L. LIONS, Some properties of viscosity solutions of Hamilton-Jacobi equations. Trans. Amer. Math. Soc., 1984.

[14] M.G. CRANDALL et P.L. LIONS, Conditions d'unicité pour les solutions généralisées des équations de Hamilton-Jacobi du premier ordre. Comptes-Rendus Acad. Sci. Paris, 292(1981), p. 183-186.

[15] M.G. CRANDALL et P.L. LIONS, Viscosity solutions of Hamilton-Jacobi equations. Trans. Amer. Math. Soc., 277 (1983), p. 1-42.

[16] M.G. CRANDALL et P.L. LIONS, Two approximations of solutions of Hamilton-Jacobi equations. Maths. Comp., à paraître.

[17] M.G. CRANDALL et P.L. LIONS : Solutions de viscosité non bornées des équations de Hamilton-Jacobi du premier ordre. Comptes-rendus Acad. Sci. Paris, 1984.

[18] M. G. CRANDALL and P.E. SOUGANIDIS, In "Procedings International Conference in Differential Equations, Alabama, 1983"

[19] A. DOUGLIS, The continuous dependance of generalized solutions of non-linear partial differential equations upon initial data. Comm. Pure Appl. Math., 14 (1961), p. 267-284.

[20] L.C. EVANS et H. ISHII, Differential games and non linear first order PDE on bounded domains, à paraître.

[21] L.C. EVANS et H. ISHII, A PDE approach to some asymptotic problems concerning random differential equations with small noise intensities , à paraître.

[22] L.C. EVANS et P.E. SOUGANIDIS, Differential games and representation formulas for solutions of Hamilton-Jacobi-Isaacs'equations. Ind. Univ. Math. J., à paraître.

[23] W.H. FLEMING, The Cauchy problem for a non linear first order partial differential equation. J. Diff. Eq., 5 (1969), p. 515-530.

[24] W.H. FLEMING, The Cauchy problem for degenerate parabolic equations, J. Math. Mech., 13 (1964), p. 987-1008.

[25] W.H. FLEMING et P.E. SOUGANIDIS, à paraître.

[26] A. FRIEDMAN, The Cauchy problem for first-order partial differential equations. Ind. Univ. Math. J., 23 (1973), p. 27-40.

[27] H. ISHII, Uniqueness of unbounded viscosity solutions of Hamilton-Jacobi equations. A paraître dans Ind. Univ. Math. J.

[28] R. JENSEN et P.L. LIONS, Some asymptotic problems in fully nonlinear elliptic equations and stochastic control. Ann. Sc. Norm. Sup. Pisa, 1984.

[29] S.N. KRUZKOV, Generalized solutions of the Hamilton-Jacobi equations of Eikonal type I. Math . USSR Sbornik, 27 (1975), p. 406-446.

[30] S.N. KRUZKOV, Generalized solutions of non-linear first order equations and of certain quasilinear parabolic equations. Vestnik Moskov Univ. Ser. I. Math Mech, 6 (1964), p. 65-74 (en Russe).

[31] S.N. KRUZKOV, Generalized solutions of nonlinear first order equations with several variables. I, II Mat. Sb, 70 (1966), p. 394-415 (en russe), 1 (1967), p. 93-116.

[32] P.L. LIONS, Generalized solutions of Hamilton-Jacobi equations. Pitman, London, 1982.

[33] P.L. LIONS, Existence results for first-order Hamilton-Jacobi equations. Ricerche di Mat., 32 (1983), p. 3-23.

[34] P.L. LIONS, On Hamilton-Jacobi semigroups. Preprint.

[35] P.L. LIONS, Optimal control of diffusion processes and Hamilton-Jacobi-Bellman equations. Part. 2, Comm. P.D.E., 8 (1983), p.1229-1276.

[36] P.L. LIONS, Solutions de viscosité des équations elliptiques du second-ordre complètement non-linéaires. In "Colloque Laurent SCHWARTZ, Ecole Polytechnique, 1983", Astérisque, 1984.

[37] P.L. LIONS, Some recent results in the optimal control of diffusion processes. Stochastic Analysis, Proceedings of the Taniguchi International Symposium on Stochastic Analysis, Katata and Kyoto, 1982. Kinokuniya , Tokyo, 1984.

[38] P.L. LIONS, In Seminaire Goulaouic-Meyer-Schwartz 1983-1984. Ecole Polytechnique, Palaiseau, 1984.

[39] P.L. LIONS et M. NISIO, A uniqueness result for the semi-group associated with the Hamilton-Jacobi-Bellmann operator. Proc. Japon Acad., 58 (1982), p. 273-276.

[40] P.L. LIONS, G. PAPANICOLAOU et S.R.S. VARADHAN, à paraître

[41] P.L. LIONS et J.C. ROCHET, Hopf formula and multi-time Hamilton-Jacobi equations. Preprint.

[42] P.L. LIONS et P.E. SOUGANIDIS, Differential games, optimal control and directional derivatives of viscosity solutions of Bellman's and Isaacs'equations. A praître dans SIAM J. Control Optim.

[43] P.E. SOUGANIDIS, Existence of viscosity solutions of Hamilton- Jacobi equations. J. Diff. Eq., à paraître.

[44] P.E. SOUGANIDIS, PhD Thesis, Univ. of Wisconsin-Madison, 1983.

Pierre Louis LIONS
Ceremade
Université Paris-Dauphine
Place de Lattre de Tassigny
75-775 - PARIS - Cedex 16

A MARINO

Critical points and operators with lack of monotonicity

INTRODUCTION

In this paper we present some ideas and some results concerning the evolution equation

(★) $0 \in U'(t) + A(U(t))$

where the operator A is defined on a Hilbert space.

The theory of monotone operators [3] , [20] has proved to be a very good framework for the study of equation (★) and has made it possible to tackle successfully several problems which do not present the classical regularity conditions. In papers [5] , [7] , [10] , [11] , [13] , [14] , we try to make a further step which, in our opinion, follows in a natural way the one made by the theory of monotone operators. We try to outline a general framework, including some significant cases which are in some sense outside the scope of the theory of monotone operators.

In § 1 we present some typical results of a variational theory (see [5] , [7] , [10] , [11] , [13]). By the operator A we mean the subdifferential (defined in a suitable way) of an extended real function f defined on a Hilbert space H and, in this case, U is the "curve of maximal slope" for f (see definition 1.2).

As in the convex case, f may assume also the value $+\infty$: in this way, for instance, we can study f on the "constraint" V (where V is a subset of H), by putting $f = +\infty$ outside V . A case in which we have applied this theory is presented in § 2 .

An important role, in the variational theory and in the problem of § 2 , is played by the notion of a point which is critical from below (see definition 1.1). To estimate the number of such points, we present a proposition which extends the classical one, due to LUSTERNIK and LEHNIRELMAN [23] , to the situations we are interested in. Another problem which can be tackled by these techniques is that of geodesics with respect to an obstacle [17] .

In § 3 we present some results of a nonvariational version of our theory (see [5] , [10] , [11] , [14]).

Finally, we show in § 4 how some first order hyperbolic systems, which are typically nonvariational, can be studied by these methods.

§ 1 - The variational case

In this section we present some definitions and some results taken from the theory developed in [5] , [7] , [10] , [11] , [13] .

In the following, H will denote a real Hilbert space, whose scalar product and norm are denoted by $(\cdot|\cdot)$ and $|\cdot|$.

We shall consider an open subset Ω of H and a lower semicontinuous function

$$f : \Omega \longrightarrow \mathbb{R} \cup \{+\infty\} .$$

We set $D(f) = \{ u \in \Omega : f(u) \in \mathbb{R} \}$.

1.1-Definition

For every u in D(f) we denote by $\partial^- f(u)$ the (possibly empty) set of the α in H such that

$$\liminf_{v \to u} \frac{f(v) - f(u) - (\alpha | v-u)}{|v-u|} \geqslant 0 .$$

If $u \in \Omega \setminus D(f)$, we set $\partial^- f(u) = \emptyset$. The set $\partial^- f(u)$ is called the subdifferential of f in u and the function f is said to be subdifferentiable in u if $\partial^- f(u) \neq \emptyset$.

Since the set $\partial^- f(u)$ is convex and closed, for every u such that $\partial^- f(u) \neq \emptyset$ we can denote by $\mathrm{grad}^- f(u)$ the element of $\partial^- f(u)$ having minimum norm.

Finally we say that a point u is critical from below if $0 \in \partial^- f(u)$ (i.e. $\mathrm{grad}^- f(u) = 0$).

We note that the notions we have just introduced agree with the well known ones, if the function f is Fréchet-differentiable or convex. In particular, it is clear that f is Fréchet-differentiable in u if and only if f and -f are subdifferentiable in u . Moreover, this subdifferential is contained in the one introduced by CLARKE and ROCKAFELLAR ([4] , [22]) and may be, in some cases, strictly contained.

We want also to emphasize that the condition $\partial^- f(u) \neq \emptyset$ provides, in some cases, useful information on u : for instance, if H is a function space and f belongs to a suitable class, then $\partial^- f(u) \neq \emptyset$ if and only if u has a certain degree of regularity.

The following notion generalizes that of solutions of the differential equation

$$U'(t) = -\operatorname{grad} f(U(t))$$

which is well known if f is C^1 or convex.

1.2- Definition

Let I be an interval in $\mathbb{R}$ with $\overset{\circ}{I} \neq \emptyset$. We say that $U : I \rightarrow \Omega$ is a curve of maximal slope for f , if

a) U is continuous on I and absolutely continuous on compact subsets of $\overset{\circ}{I}$;

b) $f \circ U$ is nonincreasing on I and

$$U'(t) \in -\partial^- f(U(t)) \qquad \text{a.e. on } \overset{\circ}{I} .$$

Now we consider a class of functions with which we have used the above definitions and which seem to be useful in tackling, for instance, problems with "constraints".

1.3 - Definition

Let $\varphi : \Omega \times \mathbb{R}^2 \rightarrow \mathbb{R}^+$ be a continuous function. We say that f is φ-convex if

$$f(v) \geqslant f(u) + (\alpha | v-u) - \varphi(u, f(u), |\alpha|)\, |u-v|^2$$

whenever $v, u \in \Omega$, $\partial^- f(u) \neq \emptyset$, $\alpha \in \partial^- f(u)$.

In particular, if

$$\varphi(u, x_1, x_2) \leqslant \varphi_o(u, x_1)\, (1 + |x_2|^p) ,$$

where $p \geqslant 0$, $\varphi_o : \Omega \times \mathbb{R} \rightarrow \mathbb{R}^+$, we say that f is φ-convex of order p.

We note that, in this definition, we do not explicitly require the subdifferentiability of f in any point.

We also note that the presence of $|\alpha|$ in φ means that we are not restricted to the class of functions which are the sum of a convex function with a $C^{1,1}$ one (whose subdifferential is $\mathcal{M}(\omega)$ in the sense of [20]).

As an example, we provide an elementary version of a more general result. We put forward the following definition:

1.4 - Definition

If A is a subset of H we set

$$I_A(u) = \begin{cases} 0 & \text{if} \quad u \in A \\ +\infty & \text{if} \quad u \in H \setminus A . \end{cases}$$

If B is another subset of H , we say that A and B are not tangential if for every u in $A \cap B$ we have

$$\partial^- I_A(u) \cap \partial^- I_B(u) = \{0\} .$$

1.5 - Proposition

Let $f_o : H \longrightarrow \mathbb{R} \cup \{+\infty\}$ be a lower semicontinuous function such that there exists q in $\mathbb{R}$ with

$$f_o(v) \geqslant f_o(u) + (\alpha | v - u) - q\, | v - u |^2$$

whenever $v, u \in H$, $\alpha \in \partial^- f(u)$.

Let M be a $C^{1,1}$ hypersurface in H such that M and $D(f_o)$ are not tangential.

Then

a) for every u in $D(f_o) \cap M$ with $\partial^-(f_o + I_M)(u) \neq \emptyset$ we have $\partial^- f_o(u) \neq \emptyset$ and

$$\partial^-(f_o + I_M)(u) = \partial^- f_o(u) + \partial^- I_M(u) ;$$

b) the function $f = f_o + I_M$ is φ-convex of order one ;

c) in particular (Lagrange theorem), if u is critical from below for f_o on M (i.e. for $f_o + I_M$), then there exists λ in $\mathbb{R}$ such that

$$\lambda \nu(u) \in \partial^- f_o(u)$$

where $\nu(u)$ is the normal to M in u .

φ-convex functions have several properties some of which are demonstrated in the two following theorems :

1.6 - Theorem

Let f be a φ-convex function. Then for every u in Ω with $\partial^- f(u) \neq \emptyset$ there exists a unique Lipschitz continuous curve of maximal slope

$$U : [0,T[\longrightarrow \Omega \quad \text{such that} \quad U(0) = u .$$

Moreover we have

$$U'_+(t) = - \operatorname{grad}^- f(U(t)),$$

$$f \circ U(t_2) - f \circ U(t_1) = - \int_{t_1}^{t_2} |\operatorname{grad}^- f(U(t))|^2 dt$$

whenever $t, t_1, t_2 \in [0,T[$.

Finally, if $(u_n)_n$ is a sequence converging to u in Ω with $\partial^- f(u_n) \neq \emptyset$, $\sup_n f(u_n) < +\infty$, $\sup_n |\operatorname{grad}^- f(u_n)| < +\infty$, we have that $\partial^- f(u) \neq \emptyset$ and there exists $T > 0$ such that the curves of maximal slope U_n, U such that $U_n(0) = u_n$, $U(0) = u$ are eventually defined on $[0,T[$ and $(U_n)_n$ converges to U uniformly on $[0,T[$.

1.7 - Theorem

Let f be a φ-convex function of order two. Then for every u in Ω with $f(u) < +\infty$ there exists a unique curve of maximal slope $U : [0,T[\longrightarrow \Omega$ such that $U(0) = u$.

Moreover we have $U'_+(t) = - \operatorname{grad}^- f(U(t))$,

$$f \circ U(t_2) - f \circ U(t_1) = - \int_{t_1}^{t_2} |\operatorname{grad}^- f(U(t))|^2 dt$$

whenever $t \in]0,T[$, $t_1, t_2 \in [0,T[$.

Finally, if $(u_n)_n$ is a sequence converging to u in Ω with $\sup_n f(u_n) < +\infty$, we have that $f(u) < +\infty$ and there exists $T > 0$ such that the curves of maximal slope U_n, U such that $U_n(0) = u_n$, $U(0) = u$ are eventually defined on $[0,T[$ and $(U_n)_n$ converges to U uniformly on $[0,T[$.

As we hinted in the introduction, in order to prove the multiplicity theorem 2.6 we have used, besides theorem 1.7, the following which extends the famous theorem due to Lusternik and Schnirelman.

Let $f : H \to \mathbb{R} \cup \{+\infty\}$ be a lower semicontinuous function. In $D(f)$ we introduce the "graph metric"

$$d^{\star}(u,v) = |u-v| + |f(u)-f(v)|$$

and we denote by $D(f)^{\star}$ the metric space $(D(f), d^{\star})$.

1.8 - Theorem

We suppose that

a) f is bounded from below ;

b) for every u in $D(f)$ there is a unique curve of maximal slope $\Phi(u,\cdot) : [0,+\infty[\longrightarrow H$ such that $\Phi(u,0) = u$ and such that $\Phi : D(f)^{\star} \times [0,+\infty[\longrightarrow D(f)^{\star}$ is continuous (this is true if, for instance, f is φ-convex of order two) ;

c) (Palais-Smale type hypothesis) if $c \in f(D(f))$ and if $(u_n)_n$ is a sequence with $\partial^- f(u_n) \neq \emptyset$, $f(u_n) \leq c$, $\lim_n |\mathrm{grad}^- f(u_n)| = 0$, then there exists a subsequence $(u_{n_h})_h$ which is convergent in $D(f)^{\star}$ to a point u with $\mathrm{grad}^- f(u) = 0$;

d) $D(f)^{\star}$ is locally contractible.

Then the following facts hold :

a) f has at least $\mathrm{cat}(D(f)^{\star})$ points which are critical from below ;

b) if $c \in f(D(f))$, we have

$$\mathrm{cat}\,\{u \in D(f) : f(u) \leq c\} < +\infty ;$$

c) if $\mathrm{cat}(D(f)^{\star}) = +\infty$, we have

$$\sup\{f(u) : u \in D(f),\ 0 \in \partial^- f(u)\} = \sup f(D(f)) \notin f(D(f)).$$

The category is evaluated in $D(f)^{\star}$.

§ 2 - Eigenvalues, with respect to an obstacle, for an elliptic operator and associated evolution equation

Here we present a problem in which we have used the techniques we have mentioned in the preceding section.

For details and proofs, see [19].

Let Ω be an open bounded subset of $\mathbb{R}^n$ and let φ_1, φ_2 be two functions defined on $\overline{\Omega}$, such that

$$\varphi_1 \leqslant \varphi_2 \text{ on } \Omega \;, \quad \varphi_1 \leqslant 0 \leqslant \varphi_2 \text{ on } \partial\Omega .$$

We set

$$K = \{u \in H_o^1(\Omega) : \varphi_1 \leqslant u \leqslant \varphi_2\}, \; S_\rho = \{u \in H_o^1(\Omega) : \int_\Omega u^2\, dx = \rho^2\} \quad (\rho > 0).$$

Finally, let G be a real function.

We want to consider the functional

$$f(u) = \begin{cases} \int_\Omega [\frac{1}{2}|Du|^2 + G(u)]dx \;, & \text{if } u \text{ belongs to the "constraint" } K \cap S_\rho \\ +\infty & \text{, if } u \text{ belongs to } L^2(\Omega) \setminus (K \cap S_\rho). \end{cases}$$

In particular, we are interested in evaluating the number of points which are critical from below (the minimum is studied, for instance, in [2]).

In the classical case, namely without the "obstacles" φ_1 and φ_2, and under the hypothesis $G \equiv 0$ the problem is well known and can be studied by techniques which are typical in linear analysis.

In contrast, the presence of the obstacles takes away "linearity" from the problem, even if $G \equiv 0$, and forces us to use techniques which are typical in nonlinear analysis. These techniques are based on the study of the topological type of $K \cap S_\rho$ and on the study in $K \cap S_\rho$ of the equation

$$U'(t) = -\, \text{grad}^- f(U(t)).$$

2.1 We make the following assumptions :

- Let φ_1, φ_2 be in $H^2(\Omega) \cap C^o(\Omega)$, with $\varphi_1 \leqslant \varphi_2$ on Ω ; if $u_K = (\varphi_1 \vee o) + (\varphi_2 \wedge 0)$, let us suppose that $u_K \in H_o^1(\Omega)$ (i.e. $\varphi_1 \leqslant 0 \leqslant \varphi_2$ on $\partial\Omega$) ;

- suppose that

$\Omega\setminus\{x : \varphi_1(x) = u_K(x)\}$, $\Omega\setminus\{x : u_K(x) = \varphi_2(x)\}$ are connected (if, for instance, $\varphi_1 < 0$ and $\varphi_2 > 0$ on Ω, we suppose Ω to be connected);

- let ρ be such that

$\inf\{\|v\|_{L^2} : v \in K\} < \rho < \sup\{\|v\|_{L^2} : v \in K\}$,

$\rho \neq \|\varphi_1\|_{L^2}$, $\rho \neq \|\varphi_2\|_{L^2}$;

- let G be a function of class C^1 such that $g = G'$ is Lipschitz continuous (for sake of simplicity)

Put $H = L^2(\Omega)$.

We note that the second and the third of the above conditions ensure that K and S_ρ are not tangential (see definition 1.4).

By the following theorem we characterize the points which are critical from below for f.

2.2. <u>Theorem</u>

Let u be in $K \cap S_\rho$. If (2.1) holds, we have that

a) f is subdifferentiable at u (in the Hilbert space $L^2(\Omega)$) if and only if $u \in H^2(\Omega)$;
moreover, if $\alpha \in \partial^- f(u)$, there exists λ in $\mathbb{R}$ such that

$$(2.2.1)\quad \begin{cases} \alpha + \lambda u + \Delta u - g(u) \leqslant 0 & \text{where } \varphi_1(x) = u(x) < \varphi_2(x) \\ \alpha + \lambda u + \Delta u - g(u) = 0 & \text{where } \varphi_1(x) < u(x) < \varphi_2(x) \\ \alpha + \lambda u + \Delta u - g(u) \geqslant 0 & \text{where } \varphi_1(x) < u(x) = \varphi_2(x) \\ \text{no condition} & \text{where } \varphi_1(x) = u(x) = \varphi_2(x) \end{cases}$$

b) in particular, u is critical from below for f if and only if $u \in H^2(\Omega)$ and there exists λ in $\mathbb{R}$ which makes (2.2.1) true with $\alpha = 0$,

i.e. if and only if there exists λ such that

$$(2.2.2)\quad \int_\Omega [DuD(v-u) + g(u)(v-u)]dx \geqslant \int_\Omega u(v-u)dx \qquad \forall v \in K;$$

c) therefore, if u is a minimum point for f on $K \cap S_\rho$ (such a point exists by a compactness argument), (2.2.1) holds with $\alpha = 0$ (see, for instance, [2]).

2.3. Theorem

If assumptions 2.1 hold, we have that

a) if I is an open interval in $\mathbb{R}$, $U : I \to K \cap S_\rho$ is a curve of maximal slope for f if and only if $U \in H^{1,2}_{loc}(I ; L^2(\Omega))$ and for every t in I $U(t) \in H^2(\Omega)$, there exists the right derivative $U'_+(t)$ and there exists $\lambda(t)$ in $\mathbb{R}$ such that

$$U'_+(t) = \begin{cases} [\lambda(t)u + \Delta u - g(u)]^+ & \text{where } \varphi_1(x) = u(x) < \varphi_2(x) \\ \lambda(t)u + \Delta u - g(u) & \text{where } \varphi_1(x) < u(x) < \varphi_2(x) \\ -[\lambda(t)u + \Delta u - g(u)]^- & \text{where } \varphi_1(x) < u(x) = \varphi_2(x) \\ 0 & \text{where } \varphi_1(x) = u(x) = \varphi_2(x) \end{cases}$$

b) f is φ-convex of order one; in particular for every u_o in $K \cap S_\rho$ there exists a unique curve of maximal slope $U : [0,+\infty[\to K \cap S_\rho$ such that $U(0) = u_o$;

c) if $(u_n)_n$ is a sequence converging to u in H with $\sup_n f(u_n) < +\infty$, and if U_n, U are the curves of maximal slope such that $U_n(0) = u_n$, $U(0) = u$, then $(U_n)_n$ converges uniformly to U on $[0,+\infty[$ and for every $t > 0$ $(f \circ U_n(t))_n$ converges to $f \circ U(t)$.

2.4. Theorem

Suppose that assumptions 2.1 hold and that

a) $g(-u) = -g(u)$ for every u in $\mathbb{R}$;

b) $\varphi_2(x) = -\varphi_1(x) = \varphi(x) \geq 0$ for every x in Ω and $\varphi(\bar{x}) > 0$ for some $\bar{x}$ in Ω.

Then there exist infinitely many (λ,u) with λ in $\mathbb{R}$ and u in $K \cap S_\rho$ solving (2.2.2.).

Such λ's are uniformly bounded from below and have a least upper bound equal to $+\infty$.

§ 3 - The nonvariational case

In this section we mention some typical results of an evolution theory for operators which are not subdifferentials of a function (see [5], [10], [11], [14]).

We shall consider, besides the lower semicontinuous function $f : \Omega \to \mathbb{R} \cup \{+\infty\}$, also an operator

$$A : H \to \mathscr{P}(H) \quad \text{such that}$$

$$D(A) = \{u \in H : A(u) \neq \emptyset\} \subset D(f).$$

We set

$$|A_o(u)| = \begin{cases} \inf\{|\alpha| : \alpha \in A(u)\}, & \text{if } u \in D(A) \\ +\infty, & \text{if } u \in H \setminus D(A). \end{cases}$$

3.1 - Definition

If $\varphi : \Omega \times \mathbb{R}^2 \to \mathbb{R}^+$ is a continuous function the operator A is said to be (φ,f)-monotone, if

$$(\alpha-\beta \mid u-v) \geq -[\varphi(u,f(u),|\alpha|) + \varphi(v,f(v),|\beta|)] \quad |u-v|^2$$

whenever $u,v \in D(A)$, $\alpha \in A(u)$, $\beta \in A(v)$.

3.2 - Definition

The operator A is said to be f-solvable at a point u of $D(A)$ if there exist

$$C_o > f(u),\ C_1 > |A_o(u)|,\ M \geq 0,\ r > 0,\ \lambda_o > 0$$

such that

a) $\forall \lambda$ in $]0,\lambda_o]$, $\forall v$ in $B(u,r) \cap D(A)$ with $f(v) \leq C_o$, $|A_o(v)| \leq C_1$, there exists w in $D(A)$ such that

$$\frac{v-w}{\lambda} \in A(w),\ \frac{|v-w|}{\lambda} \leq M,\ f(w) \leq f(v) + \lambda M;$$

b) for every sequence $(v_n)_n$ in $D(A)$ converging to v in $B(u,r)$ with $f(v_n) \leq C_o$, for every sequence $(\alpha_n)_n$ weakly convergent to α with $\alpha_n \in A(v_n)$, $|\alpha_n| \leq C_1$, we have $v \in D(A)$ and $\alpha \in A(v)$.

3.3 - Remark

If f is a φ-convex function, then $\partial^- f$ is a (φ,f)-monotone operator which is f-solvable at every point of $D(A)$.

If A is a Lipschitz continuous perturbation of a maximal monotone operator (see [3], [20]), and if we take $f \equiv 0$, then A is a (φ,f)-monotone operator, for a suitable constant function φ, which is f-solvable at every point of $D(A)$.

3.4 - Proposition

Let A be a (φ,f)-monotone operator which is f-solvable at a point u of $D(A)$. Then $A(u)$ has a unique element of minimal norm that we denote by $A_o(u)$.

3.5 - Definition

Let I be an interval in $\mathbb{R}$ with $\mathring{I} \neq \emptyset$. We say that $U : I \to \Omega$ is an evolution curve for A, if

a) U is continuous on I and absolutely continuous on compact subsets of $\mathring{I}$;

b) $f \circ U$ is bounded on compact subsets of $\mathring{I}$ and

$$U'(t) \in - A(U(t)) \qquad \text{a.e. on } \mathring{I}.$$

3.6 - Theorem

Let A be a (φ,f)-monotone operator which is f-solvable at every point of $D(A)$. Then for every u in $D(A)$ there exists a unique Lipschitz continuous evolution curve $U : [0,T[\longrightarrow D(A)$ such that $U(0) = u$ and such that $f \circ U$ is bounded on $[0,T[$.

Moreover

$$U'_+(t) = - A_o(U(t)) \qquad \forall\, t \quad \text{in} \quad [0,T[\,.$$

Finally, if $(u_n)_n$ is a sequence in $D(A)$ converging to u in Ω with $\sup_n |A_o(u_n)| < +\infty$, $\sup_n f(u_n) < +\infty$, we have that $u \in D(A)$ and there exists $T > 0$ such that the evolution curves U_n, U such that $U_n(0) = u_n$, $U(0) = u$ are eventually defined on $[0,T[$ and $(U_n)_n$ converges to U uniformly on $[0,T[$.

§ 4 - Some first-order hyperbolic systems

In this section we show, as an example, that some first-order nonlinear hyperbolic systems can be treated by the techniques exposed in § 3 .

4.1 - Problem

More precisely, we consider :

Find $u :]-T,T[\times \mathbb{R}^n \to \mathbb{R}^N$ solving

$$(4.1.1.)\quad \begin{cases} \dfrac{\partial u}{\partial t} + F(x,u,\dfrac{\partial u}{\partial x}) = 0 \\ u(0,x) = u_o(x) \end{cases}$$

in the case

$$\frac{\partial F_h}{\partial p_j^k}(x,u,p) = \frac{\partial F_k}{\partial p_j^h}(x,u,p) \qquad 1 \leqslant j \leqslant n\ ,\ 1 \leqslant h,k \leqslant N\ .$$

This kind of problem was considered in [15] in the quasi-linear case and in [8] in the nonlinear case.

According to [8] , we suppose that

$$F : \mathbb{R}^n \times \mathbb{R}^N \times \mathbb{R}^{nN} \to \mathbb{R}^N$$

is a map of class C^k with $k > \frac{n}{2} + 2$, which is periodic in every x_i of period 2π .

We denote by L^2_Q , H^k_Q the closure of

$$\{u \in C^\infty(\mathbb{R}^n ; \mathbb{R}^N) : u \text{ is periodic in every } x_i \text{ of period } 2\pi\}$$

in $L^2(Q)$ (resp. $H^k(Q)$), where $Q =]0,2\pi[^n$.

In [8] the following result is proved :

4.2 - Theorem

For every u_o in H^k_Q there is $T > 0$ and a unique u in

$C^o(-T,T ; H^k_Q) \cap C^1(-T,T ; L^2_Q)$

solving (4.1.1.).

Here we wish to point out that this result can also be proved by means of theorem 3.6. Indeed, we can set

$$H = L^2_Q \ , \ D(A) = H^k_Q \ , \ A(u) = F(x,u,Du),$$

$$f(u) = \begin{cases} \|u\|_{H^k_Q} & \text{if} \quad u \in H^k_Q \\ +\infty & \text{if} \quad u \in L^2_Q \setminus H^k_Q . \end{cases}$$

Then, if $\varphi : \mathbb{R} \to \mathbb{R}^+$ is a suitable continuous function, we have

$$(\alpha - \beta \,|\, u - v) \geqslant - [\varphi(f(u)) + \varphi(f(v))] \ |u - v|^2$$

whenever

$$\alpha \in A(u), \quad \beta \in A(v).$$

On the other hand, A is f-solvable at every u of $D(A)$. This can be proved by means of the following result (see [9]).

4.3 - Theorem

Let $G : \mathbb{R}^n \times \mathbb{R}^N \times \mathbb{R}^{nN} \to \mathbb{R}^N$ be a map of class C^k with $k > \frac{n}{2} + 2$, which is periodic in every x_i of period 2π.

Suppose that

$$\frac{\partial G_h}{\partial p^\ell_j}(x,u,p) = \frac{\partial G_\ell}{\partial p^h_j}(x,u,p), \ 1 \leqslant j \leqslant n, \ 1 \leqslant h,\ell \leqslant N$$

and that there exists $\gamma > 0$ such that

$$(\eta | [\frac{\partial G}{\partial u}(x,0,0) - \frac{1}{2} \sum_1^n j \frac{\partial^2 G}{\partial x_j \partial p_j}(x,0,0)] \eta) \geqslant \gamma |\eta|^2$$

$$k \sum_{1\,i,j}^n \xi_i \ \xi_j \ (\eta | \frac{\partial^2 G}{\partial x_i \partial p_j}(x,0,0)\eta) +$$

$$+ |\xi|^2 (\eta | [\frac{\partial G}{\partial u}(x,0,0) - \frac{1}{2} \sum_1^n j \frac{\partial^2 G}{\partial x_j \partial p_j}(x,0,0)]\eta) \geqslant \gamma |\xi|^2 \ |\eta|^2$$

for every $x_i \ \xi$ in $\mathbb{R}^n$, η in $\mathbb{R}^N$.

Then there exists $\varepsilon_o > 0$ such that for every ε in $]0,\varepsilon_o]$ the problem

$$\begin{cases} G(x,u,Du) = G(x,0,0) + g(x) \\ u \in H^k_Q , \ \|u\|_{H^k_Q} < \varepsilon \end{cases}$$

has a unique solution, if $g \in H^k_Q$, $\|g\|_{H^k_Q} < \delta$ and $\delta > 0$ is sufficiently small.

REFERENCES

[1] V. BENCI : Positive solutions of some eigenvalue problems in the theory of variational inequalities, J. Mat. Anal. Appl. 61 (1977), 165-187.

[2] V. BENCI - A. MICHELETTI : Su un problema di autovalori per disequazioni variazionali, Ann. Mat. Pura Appl. (4) 107 (1976) 359-371.

[3] H. BREZIS : Opérateurs maximaux monotones, Notes de Mathematica (50), North-Holland 1973.

[4] F.H. CLARKE : Optimization and non-smooth analysis, Wiley 1983.

[5] E. DE GIORGI - M. DEGIOVANNI - A. MARINO - M. TOSQUES : Evolution equations for a class of nonlinear operators, Atti Accad. Naz. Lincei, Rend. Cl. Sci. Fis. Mat. Natur. (8) 75 (1983), 1-8.

[6] E. DE GIORGI - M. DEGIOVANNI - M. TOSQUES : Recenti sviluppi della Γ-convergenza in problemi ellittici, parabolici ed iperbolici, Proceedings of the International Meeting dedicated to the memory of professor Carlo MIRANDA, Liguori 1982.

[7] E. DE GIORGI - A. MARINO - M. TOSQUES : Funzioni (p,q)-convesse, Atti. Accad. Naz. Lincei, Rend. Cl. Sci. Fis. Mat. Natur. (8) 73 (1982), 6-14.

[8] M. DEGIOVANNI : Alcuni sistemi iperbolici non lineari del primo ordine visti come problemi di evoluzione in presenza di più norme, Boll. Un. Mat. Ital. B (6) 1 (1982), 47-73.

[9] M. DEGIOVANNI - A. MARINO - S. SPAGNOLO : Alcuni sistemi differenziali non lineari del primo ordine sul toro, Boll. Un. Mat. Ital. B (6) 1 (1982), 17-45.

[10] M. DEGIOVANNI - A. MARINO - M. TOSQUES : General properties of (p,q)-convex functions and (p,q)-monotone operators, Ricerche Mat. 32 (1983), 285-319.

[11] M. DEGIOVANNI - A. MARINO - M. TOSQUES : Evolution equations associated with (p,q)-convex functions and (p,q)-monotone operators, Ricerche Mat., in press.

[12] M. DEGIOVANNI - A. MARINO - M. TOSQUES : Critical points and evolution equations, Proceedings of the International Conference on Multifunctions and Integrands, Springer Verlag, in press.

[13] M. DEGIOVANNI - A. MARINO - M. TOSQUES : Evolution equations with lack of convexity, J. Nonlinear Anal., in press.

[14] M. DEGIOVANNI - M. TOSQUES : Evolution equations for (φ,f)-monotone operators, Boll. Un. Mat. Ital. B, in press.

[15] T. KATO : The Cauchy problem for quasi-linear symmetric hyperbolic systems, Arch. Rational Mech. Anal. 58 (1975), 181-205.

[16] M. KUCERA - J. NECAS - J. SOUCEK : The eigenvalue problem for variational inequalities and a new version of the Lusternik-Schnirelman theory, in "Nonlinear analysis", a collection of papers in honour of Enrich ROTHE, Academic Press 1978.

[17] A. MARINO - D. SCOLOZZI : Geodetiche con ostacolo, Boll. Un. Mat. Ital. B (6) 2 (1983), 1-31.

[18] A. MARINO - D. SCOLOZZI : Punti inferiormente stationari ed equazioni di evoluzione con vincoli unilaterali non convessi, Rend. Sem. Mat. Fis. Milano, in press.

[19] A. MARINO - D. SCOLOZZI : Moltiplicatoxi di Lagrange e φ-convessita submitted.

[20] A. PAZY : Semi-groups of nonlinear contractions in Hilbert space, in "Problems in nonlinear analysis", Cremonese 1971.

[21] R.C. RIDDELL : Eigenvalue problems for nonlinear elliptic variational inequalities, Nonlinear Anal. 3 (1979), 1-33.

[22] R.T. ROCKAFFELLAR : Generalized directional derivatives and subgradients of non convex functions, Canad. J. Math. 32 (1980), 257-280.

[23] J.T. SCHWARTZ : Nonlinear functional analysis, Gordon and Breach 1978.

Antonio MARINO
Dipartimento di Matematica
Via Buonarroti, 2
I 56100 - PISA

Marco DEGIOVANNI
Scuola Normale Superiore
Piazza dei Cavalieri, 7
I 56100 - PISA

M MIRANDA

Compactness of solutions to the minimal surface equation

Among the most surprising results of recent years in the theory of calculus of variations and partial differential equations is the existence of minimal singular surfaces of codimension one in the Euclidean space E^8 and the existence of non-trivial solutions to the minimal surface equation in the whole R^8. These two facts were established by E. BOMBIERI, E. DE GIORGI and E. GIUSTI in 1969. What is even more surprising is that the techniques used by these authors were the same as were used to prove, up to dimension seven, the results going in the opposite direction : minimal surfaces of codimension one in E^7 are analytic and the only solutions to the minimal surface equation in R^7 are polynomials of degree 0 or 1 . The aim of this paper is to show how the results of BOMBIERI-DE GIORGI-GIUSTI can easily be derived from what was known about minimal surfaces before 1969. For this purpose the following remark is fundamental : "the set of all solutions to the minimal surface equations is compact".

Naturally this statement must be made more precise.

Compactness, with respect to which convergence ?

Solutions, in what sense ?

It is obvious that a sequence of solutions may tend to $+\infty$ somewhere and to $-\infty$ somewhere else, or to a classical solution in one subregion and to $+\infty$, $-\infty$ in some others.

A general compactness theorem such as the one stated above needs a convenient generalization of the definition of solutions.

I presented such a generalization in an article published in a special issue of the Annali della Scuola Normale di Pisa, dedicated to Jean LERAY. A generalized solution to the minimal surface equation, in the open set Ω of $\mathbb{R}^n$, is defined as being a Lebesgue-measurable function f with values in $\mathbb{R} \cup \{-\infty, +\infty\}$, such that the set

$$E = \{(x,t) \mid x \in \Omega, t \in \mathbb{R}, t < f(x)\}$$

has minimal boundary in $\Omega \times R$, according to the definition introduced by E. DE GIORGI in 1960 (see [2]) .

The compactness of these solutions is quite natural, being nothing else than the compactness of a class of characteristic functions whose gradients, in the sense of distributions, are locally bounded.

A more convenient and precise formulation of the compactness theorem can be given, allowing the domains of the solutions to change :

Compactness Theorem

"Let $\{\Omega_j\}_j$ be an increasing sequence of open sets of R^n. Let f_j be a generalized solution to the minimal surface equation in Ω_j , for all j . There exists a strictly increasing sequence of integers $\{j(k)\}_k$ and a generalized solution f defined in $\Omega = \cup_j \Omega_j$, such that

$$\lim_k f_{j(k)} = f \text{ , almost everywhere in } \Omega \text{".}$$

Let us now return to our problem. Consider the 4th degree polynomial

$$P(x,y) = |x|^4 - |y|^4 \text{ , where } x \in R^k \text{ , } y \in R^k .$$

If P were a solution to the minimal surface equation, one would find a non-trivial solution in R^{2k} and, at the same time, the sequence of solutions

$$P_j(x,y) = j^{-1} P(jx,jy) = j^3 P(x,y),$$

diverging to $+\infty$ if $|x| > |y|$ and to $-\infty$ if $|x| < |y|$, would prove that the cylinder

$$\{(x,y,t) \mid x \in R^k \text{ , } y \in R^k \text{ , } t \in R, |x| = |y| \}$$

is a minimal boundary, therefore its cross section

$$\{(x,y) \mid x \in R^k, y \in R^k \text{ , } |x| = |y| \}$$

would be a minimal singular boundary.

But we know that this is impossible if $k < 4$. What we can see, by means of a an elementary computation, is that

$$MP(x,y) = \text{div}\left(\frac{\text{grad}(x,y)}{\sqrt{1+|\text{grad}P(x,y)|^2}}\right) =$$

$$= \frac{(4n+8)\{1+16(|x|^6 + |y|^6)\} - 12 . 16(|x|^2 + |y|^2)(|x|^4 + |y|^4)}{\sqrt{1 + 16(|x|^6 + |y|^6)}} \cdot (|x|^2 - |y|^2)$$

which implies, for $k \geqslant 4$,

$$MP(x,y) > 0 \quad , \quad \text{if} \quad |x| > |y| \quad ,$$

$$MP(x,y) < 0 \quad , \quad \text{if} \quad |x| < |y| \quad .$$

Therefore, for $k \geqslant 4$, P is a subsolution to the minimal surface equation where it is positive and a supersolution where it is negative. From this can reach the same conclusions, as if we had $MP = 0$.

1 - The surface

$$\{(x,y) \mid x \in R^k, \ y \in R^k, \ |x| = |y|\}$$

is minimal if $k \geqslant 4$.

Proof

Consider the sequence $\{f_j\}_j$ of solutions to the minimal surface equation defined in the sequence of balls

$$B_j = \{(x,y) \mid x \in R^k, \ |y| \in R^k, \ |x|^2 + |y|^2 < j^2\} \ ,$$

and satisfying

$$f_j = P_j \quad \text{on} \quad \partial B_j .$$

The Maximum Principle applied to the pair (f_j, P_j) gives

$$f_j(x,y) \geqslant P_j(x,y) \quad \text{if} \quad |x| > |y| \quad \text{and} \quad |x|^2 + |y|^2 < j^2 \ ,$$

$$f_j(x,y) \leqslant P_j(x,y) \quad \text{if} \quad |x| < |y| \quad \text{and} \quad |x|^2 + |y|^2 < j^2 \ .$$

Then

$$\lim_j f_j(x,y) = +\infty \ , \quad \text{if} \quad |x| > |y| \ ,$$

$$\lim_j f_j(x,y) = -\infty \ , \quad \text{if} \quad |x| < |y| \ ,$$

which is sufficient to prove, as we have already remarked, that the cone

$$\{(x,y) \mid x \in R^k, y \in R^k, |x| = |y|\}$$

is a minimal surface.

2 - Existence of non trivial solutions to the minimal surface equation in R^n for $n \geq 8$.

<u>Proof</u>

Consider the sequence $\{g_j\}_j$ of solutions to the minimal surface equation defined in the sequence of balls B_j and satisfying $g_j = P$ on ∂B_j.

We get again, from the maximum principle,

$$g_j(x,y) \geq P(x,y) \quad \text{if} \quad |x| > |y| \quad \text{and} \quad |x|^2 + |y|^2 < j^2 ,$$

$$g_j(x,y) \leq P(x,y) \quad \text{if} \quad |x| < |y| \quad \text{and} \quad |x|^2 + |y|^2 < j^2 .$$

Now, see [1], for $k \geq 4$ there exists a function

$$Q : R^{2k} \to R$$

such that

$$Q(x,y) \geq P(x,y) , \quad MQ(x,y) \leq 0 \quad \text{if} \quad |x| > |y|$$

$$Q(x,y) \leq P(x,y) , \quad MQ(x,y) \geq 0 \quad \text{if} \quad |x| < |y| .$$

Therefore, from the maximum principle, we obtain

$$P(x,y) \leq g_j(x,y) \leq Q(x,y) \quad \text{if} \quad |x| > |y| \quad \text{and} \quad |x|^2 + |y|^2 < j^2 ,$$

$$Q(x,y) \leq g_j(x,y) \leq P(x,y) \quad \text{if} \quad |x| < |y| \quad \text{and} \quad |x|^2 + |y|^2 < j^2 .$$

These inequalities and our compactness theorem easily imply the existence of a non-trivial solution to the minimal surfaces equation in R^{2k}, if $k \geq 4$.

REFERENCES

[1] E. BOMBIERI, E. DE GIORGI, E. GIUSTI - Minimal Cones and the Bernstein Problem, Inv. Math. 7 (1969), 243-268.

[2] E. DE GIORGI, F. COLOMBINI, L.C. PICCININI - Frontiere orientate di misura minima e questioni collegate, Pubbl. Cl. di Scienze, Scuola Norm. Sup. Pisa, 1972.

[3] M. MIRANDA - Superficie minime illimitate, Ann. Scuola Norm. Sup. Pisa 2 (1977), 313-322.

[4] U. MASSARI, M. MIRANDA - A Remark on Minimal Cones, Boll. Unione Mat. Ital. 2-A (1983), 123-125.

Mario MIRANDA
Dipartimento di Matematica
Università di Trento
I - 38050 POVO (TN)
(Italy).

L MODICA

Gamma convergence and stochastic homogenization

The present paper which reports joint work with Gianni DAL MASO [1] shows how certain results on nonlinear stochastic homogenization can be obtained by combining some of E.de Giorgi's methods of Γ-convergence with simple techniques of Probability Theory.

As is well known, the Γ-convergence theory began with E. de Giorgi's and S. Spagnolo's theorem (1974) asserting the convergence of the energy functionals for a G-converging sequence (in the meaning of Spagnolo) of second order and elliptic differential operators. Since the main example of G-convergence is given by an homogenized variational elliptic equation, the fact that one of the first examples of Γ-convergence is a sequence of integral functionals is natural ; this concerns nonlinear homogenization since the Euler equation of a nonquadratic functional is nonlinear.

The following general P. Marcellini's theorem (1978) is recalled here since this theorem is a model for our theorem, in that our thoerem can be viewed as a stochastic version of P. Marcellini's theorem,

Theorem [2]

Let $f = f(x,p) : \mathbb{R}^n \times \mathbb{R}^n \to \mathbb{R}$ be a function measurable and periodic in x , convex with respect to p and such that

$$c_1|p|^\alpha \leqslant f(x,p) \leqslant c_2(1 + |P|^\alpha \quad \text{for all } x \text{ and } p$$

where constants $\alpha > 1$, $c_2 \geqslant c_1 > 0$ are given. Denoting $\mathcal{a}_o$ the family of all open and bounded subsets of $\mathbb{R}^n$, F_ε denotes for any $\varepsilon > 0$ the following functional on $L^\alpha_{loc}(\mathbb{R}^n) \times \mathcal{a}_o$

$$F_\varepsilon(u,A) = \begin{cases} \int_A f(x/\varepsilon, \nabla u(x)dx & \text{if } u|_A \in W^{1,\alpha}(A) \\ +\infty & \text{otherwise.} \end{cases}$$

Hence for any fixed $A \in \mathcal{O}_0$, the sequence $(F_\varepsilon(.,A))$ is $\Gamma(L^\alpha(A)^-)$ -converging for $\varepsilon \to 0$ to the following functional F_0

$$F_0(u,A) = \begin{cases} \int_A f_0(\nabla u(x))\, dx & \text{if } u|_A \in W^{1,\alpha}(A) \\ +\infty & \text{otherwise} \end{cases}$$

where :

$$f_0(p) = \lim_{t\to\infty} \min_u \{|Q_t|^{-1} F_1(u,Q_t) ; u - \ell_p \in W_0^{1,\alpha}(Q_t)\}.$$

In this last formula $|Q_t|$ denotes the Lebesgue-measure of the hyper-cube Q_t of $\mathbb{R}^n$ where $|x_i| < t$ for $i = 1,\dots,n$; and $\ell_p(x) = p - x$ denotes the linear function on $\mathbb{R}^n$ with gradient $\nabla\ell_p = p$. We next consider stochastic homogenization. In the simplest case of deterministic homogenization the physical meaning concerns the limit electrostatic or thermal behaviour of a material with periodic structure consisting of superposed and alternated layers of which the thickness tends to zero. Hence for the elaboration of the corresponding stochastic theory, its seems quite natural to think about a random superposition of layers.

As for the deterministic case, the first examples of stochastic homogenization concerned second order elliptic equations and were given by V.V. Yurinskij [3] , S.M. Kozlov [4] , G.E. Papanicolaou and S.R.S. Varadhan [5] . The first two of these authors used Γ-convergence, the last two used the multiscale theory, and all four used the ergodic theory.

One result given by these Authors is the following

<u>Theorem [4]</u>

The following Dirichlet problems are considered :

$$(P_\varepsilon) \quad \begin{cases} \Sigma_{i,j = 1\dots n}\ \partial_i (a_{ij}(w,x/\varepsilon)\, \partial_j u) = \varphi \\ u - u_0 \in W_0^{1,2}(A) \end{cases}$$

where $\partial_i = \partial/\partial x_i$, A denotes a bounded open subset of $\mathbb{R}^n$, $\varphi \in L^2(A)$, $u^o \in W^{1,2}(A)$, where $(\Omega,\mathcal{T},P)$ denotes a probability space. The random field defined by the coefficients $a_{ij} : \Omega \times \mathbb{R}^n \to \mathbb{R}$ is assumed stochastically periodic. An ellipticity condition uniform in x and w is assumed :

$$c_1 |\xi|^2 \leqslant \Sigma_{i,j = 1\dots n} \quad a_{ij}(w,x)\, \xi_i\, \xi_j \leqslant c_2 |\xi|^2$$

with c_1 and $c_2 > 0$, $a_{ij} = a_{ji}$. Then there exist random variables a^o_{ij} satisfying an ellipticity condition uniform in w such that the solution $u_\varepsilon(w,\cdot)$ of the problems (P_ε) converge almost surely in $L^2(A)$ to the solution of the following Dirichlet problem

$$\Sigma_{ij=1\dots n} \quad a^o_{ij}(w)\, \partial_{ij}\, u = \varphi$$

$$u - u_o \in W^{1,2}_o(A)\,.$$

Moreover the random variables a^o_{ij} are constant if the random field (a_{ij}) is ergodic.

As indicated previously, this result will be extended to the nonlinear Euler-Lagrange equations corresponding to the following functionals.
Let $\mathcal{A}_o$ be the set of all bounded open subsets of $\mathbf{R}^n$. For arbitrary reals $\alpha > 1$, $c_2 \geqslant c_1 > 0, \mathcal{F} = \mathcal{F}(\alpha, c_1, c_2)$ denotes the set of all functions $F : L^\alpha_{loc}(\mathbf{R}^n) \times \mathcal{A}_o \longrightarrow \overline{\mathbf{R}} = \mathbf{R} \cup \{-\infty, +\infty\}$ of the type

$$F(u,A) = \begin{cases} \int_A f(x, \nabla u(x))\, dx & \text{if } \quad u\big|_A \in W^{1,\alpha}(A) \\ +\infty & \text{otherwise} \end{cases}$$

where $f : \mathbf{R}^n \times \mathbf{R}^n \longrightarrow \mathbf{R}$ denotes any functions such that

$$\begin{cases} c_1 |p|^\alpha \leqslant f(x,p) \leqslant c_2 (1 + |p|^\alpha) \; ; \; (x,p) \in \mathbf{R}^n \times \mathbf{R}^n \\ f(x,p) \text{ is Lebesgue-measurable in } x \text{ and convex in } p \,. \end{cases}$$

Introducing some probability space $(\Omega, \mathcal{T}, P)$ random functionals $F: \Omega \longrightarrow \mathcal{F}$ of the previous type must be defined. Hence a σ-field is needed on $\mathcal{F}$. Since $\mathcal{F} \subset \overline{\mathbf{R}}^T$ with $T = L^\alpha_{loc}(\mathbf{R}^n) \times \mathcal{A}_o$, $\mathcal{F}$ can be viewed as a topological subspace or as a measurable subspace of a product space ; hence as a candidate for this σ-field. But since the convergence of these functionals and also of their minima will be considered, a compact metrizable topology on $\mathcal{F}$ such that the minimization of functionals is continuous is useful. This topology is given by the Γ-convergence and more precisely by the following theorem :

Theorem

The Γ-convergence of a sequence (F_h) in $\mathcal{F}$ is defined by the following two conditions ;

a) For arbitrary $A \in \mathcal{Q}_0$, $u_\infty \in L^\alpha_{loc}(\mathbb{R}^n)$, and any sequence (u_h) converging in $L^\alpha_{loc}(\mathbb{R}^n)$ to u_∞ :

$$\varliminf_{h\to\infty} F_h(u_h,A) \geqslant F_\infty(u_\infty,A) .$$

b) For arbitrary $A \in \mathcal{Q}_0$, $u_\infty \in L^\alpha_{loc}(\mathbb{R}^n)$, a sequence (u_α) converging in $L^\alpha_{loc}(\mathbb{R}^n)$ to u exists such that :

$$\varlimsup_{h\to\infty} F_h(u_h,A) \leqslant F_\infty(u_\infty,A) .$$

Then the Γ-converging sequences of $\mathcal{F}$ are the converging sequences for some metric d on X such that $(\mathcal{F},d)$ is compact. Moreover, for arbitrary $A \in \mathcal{Q}_0$, $\varphi \in L^\beta(A)$ with $\alpha^{-1} + \beta^{-1} = 1$ and $u_0 \in W^{1,\alpha}(A)$, the following mapping is continuous:

$$(\mathcal{F},d) \ni F \to \mathcal{M}_{A,\varphi,\mu_0}(F) = \min_u \{ F(u,A) + \int_A \varphi u \, dx \, ; u - u_0 \in W^{1,\alpha}_0(A) \}.$$

Denoting $\mathcal{T}_B$ the Borel σ-field of $(\mathcal{F},d)$, a random functional F is defined as a measurable mapping $(\Omega, \ ,P) \longrightarrow (\mathcal{F},\mathcal{T}_B)$.

In order to define periodic random functionals and stochastic homogenization processes for arbitrary $c \in \mathbb{R}^n$ and $\varepsilon > 0$, the translation τ_c and scaling ρ_ε in $\mathbb{R}^n$:

$$\tau_c(A) = A + c \quad ; \quad (\tau_c u)(x) = u(x-c)$$

$$\rho_\varepsilon(A) = \varepsilon A \qquad (\varphi_\varepsilon u)(x) = u(x/\varepsilon)$$

are extended in the following way to random functionals :

$$(\tau_c F)(\omega) \ (u,A) = F(\omega)(\tau_c u, \tau_c A)$$

$$(\rho_\varepsilon F)(\omega) \ (u,A) = F(\omega)(\rho_\varepsilon u, \rho_\varepsilon A)$$

for arbitrary $w \in \Omega$ and $(u,A) \in T$. In particular for F of type

$$F(w)(u,A) = \int_A f(w,x,\nabla u(x))\,dx$$

for arbitrary u such that $u|_A \in W^{1,\alpha}(A)$, we can easily show:

$$\tau_c F(w)(u,A) = \int_A f(w,x+c,\nabla u(x))\,dx$$

$$\rho_\varepsilon F(w)(u,A) = \int_A f(w,x/\varepsilon,\nabla u(x))\,dx.$$

In order to prove the measurability of $\tau_c F$ and $\rho_\varepsilon F$, the following proposition is useful. The interest of this proposition lies in the fact that Γ-convergency is generally independent of the simple convergence. In particular the simple limit exists in homogenization and is different from the Γ-limit.

Proposition

Putting $T = L^\alpha_{loc}(\mathbb{R}^n) \times \mathcal{U}^o$, the product σ-field of $\overline{\mathbb{R}}^T$ induces on $\mathcal{F}$ the Borel σ-field $\mathcal{T}_B$ of $(\mathcal{F},d)$.

Definition

a) A random functional F is called stochastic periodic of period one if F and $\tau_z F$ have the same distribution for arbitrary $z \in \mathbb{Z}^n$.

b) A family (F_ε) of random functionals is called a stochastic homogenization process modelled on the random functional F if F_ε and $\rho_\varepsilon F$ have the same distribution for all $\varepsilon > 0$.

The main result is the following

Theorem

Let (F_ε) be a stochastic homogenization process modelled on the periodic random functional F. Suppose that a real M exists such that for an arbitrary pair (A,B) of open subsets such that $\text{dist}(A,B) \geq M$ the two following families of random variables are independent

$$\{F(.)(u,A)\,;\, u \in L^\alpha_{loc}(\mathbb{R}^n)\} \quad \text{and} \quad \{F(.)(u,B)\,;\, u \in L^\alpha_{loc}(\mathbb{R}^n)\}.$$

Then for $\varepsilon \to 0_+$, (F_ε) converges in probability to a random functional F_o constant on Ω of the type

$$F_o(u,A) = \begin{cases} \int_A f_o(\nabla u(x))\,dx & \text{if } u|_A \in W^{1,\alpha}(A) \\ +\infty & \text{otherwise} \end{cases}$$

where for all $p \in \mathbb{R}^n$

$$f_o(p) = \lim_{t\to\infty} \int_\Omega |Q_t|^{-1} \min_u \{F(w)(u,Q_t) \,;\, u - \ell_p \in W_o^{1,\alpha}(Q_t)\}\, dP(w)$$

where $Q_t = \{x \in \mathbb{R}^n \,;\, |x_i| \leqslant t \,;\, i = 1,\dots,n\}$, $|Q_t| = (2t)^n$ is the measure of Q_t, ℓ_p denotes the linear form on $\mathbb{R}^n$ with gradient μ.

The next corollary follows directly from the continuity of the minimization operator :

Corollary

With the hypothesis of the previous theorem, the real random variables $\mathcal{M}_{A,\varphi,u_o}(F_\varepsilon(.))$ converge in probability for $\varepsilon \to 0_+$ to $\mathcal{M}_{A,\varphi,u_o}(F_o)$.

A similar corollary is valid for the minimizing points, proving the Kozlov-Yurinskij, Papanicolavu and Varadhan theorem in the nonquadratic case. But in view of our hypothesis on the distributions, the almost sure convergence cannot be deduced.

Finally an easy example is given. For $\lambda > 0$ and $\Lambda > 0$ given with $\lambda \neq \Lambda$, the two points set $\{\lambda, \Lambda\}$ is endowed with the probability $r \in]0,1[$ on λ an $1-r$ on Λ. The set $\Omega = \{\lambda, \Lambda\}^{\mathbf{Z}}$ of double infinite sequences of reals $\in \{\lambda, \Lambda\}$ is endowed with the product probability. The following random process is defined on the line

$$w = (w_k)_{k\in\mathbf{Z}} \,;\, t \longrightarrow a(w,t) = w_k \quad \text{for} \quad t \in [k,k+1[$$

This process takes values in $\{\lambda, \Lambda\}$, is constant on all intervals $[k,k+1[$ and has period one. The following random functionals are defined

$$F(w)(u,A) = \begin{cases} \int_A a(w,t)\, |u'(t)|^\alpha\, dt & \text{if } u|_A \in W^{1,\alpha}(A) \\ +\infty & \text{otherwise} \end{cases}$$

and $F_\varepsilon = \rho_\varepsilon F$ with $\alpha > 1$ fixed. Our theorem imples for $\varepsilon \to 0_+$ the convergence in probability of (F_ε) to F_o given by

$$F_o(u,\underline{A}) = \begin{cases} a_o \int_A |u'(t)|^\alpha \, dt & \text{if } u|_A \in W^{1,\alpha}(A) \\ +\infty & \text{if not} \end{cases}$$

where $a_o = ((r)\lambda^\beta + (1-r)\Lambda^\beta)^{1/\beta}$ with $\beta = (1-\alpha)^{-1}$. The particular case $\alpha = 2$, $r = 1/2$ gives the harmonic mean of λ and Λ, i.e. as in the deterministic unidimensional homogenization theory.

REFERENCES

[1] G. DAL MASO - L. MODICA : Non linear stochastic homogenisation (to appear).

[2] P. MARCELLINI : Periodic solutions and homogenization of nonlinear variational problems. Ann. Mat. Pura Appl. (4) 117 (1978) 139-152.

[3] V.V. YURINSKIJ : Averaging an elliptic boundary-value problem with random coefficients. Siberian Math. Journal 21 (1980) 470-482.

[4] S.M. KOZLOV : Averaging of random operators. Math. USSR Sbornik,37 (1980) 167-180.

[5] G.C. PAPANICOLAOU - S.R.S. VARADHAN : Boundary value problems with rapidly oscillating random coefficients. Proc. of Colloq. on Random Fields, Rigorous results in statistical mechanics and quantum field theory, ed. by J. Fritz, J.L. Lebowitz, D. Szàsz. Colloquia Mathematica Societ. Janos Bolyai, 10, North-Holland, Amsterdam, 1979, 835-873.

Luciano MODICA
Dipartamento di Matematica
Universita di PISA

E SANCHEZ-PALENCIA

Problèmes mathématiques liés à l'écoulement d'un fluide visqueux à travers une grille

1 - INTRODUCTION

Dans cet exposé je vais parler du paradoxe de Stokes et de certains problèmes en relation avec lui. Avant de commencer l'exposé proprement dit, je voudrais dire quelques mots sur le domaine d'interaction des mathématiques et de la mécanique, domaine où s'inscrit l'oeuvre du Professeur De GIORGI. Dans ce domaine, les différences sont purement artificielles et la volonté d'avancer et de comprendre l'emporte facilement sur les différences de méthode ou d'école. Il y a certes, des différences de langage : en mathématiques, il y a des hypothèses, des lemmes, des théorèmes, en mécanique le discours est souvent plus fluide et moins précis ; il y a des principes, parfois des théorèmes et il y a des paradoxes. Or, il ne faudrait pas se méprendre, les paradoxes qui se trouvent dans la littérature sont souvent bien expliqués et pourraient bien s'appeler des propositions. On a gardé le terme de paradoxe pour signaler qu'il s'agit là d'un résultat différent de celui auquel on s'attendait. Aussi, désigner une proposition par paradoxe, c'est un peu sacrifier la logique à l'histoire et nous rappeler le cheminement lent et difficile de la pensée scientifique. Et c'est peur être le moment d'ouvrir une parenthèse pour dire, si besoin en était, que ce cheminement est inséparable du cheminement lent et difficile de l'humanité et de ses luttes de toutes sortes et de saluer encore là, l'activité remarquable du Professeur de GIORGI dans la défense des droits de l'homme.

Considérons dans le plan R^2 un domaine borné $\mathcal{O}$ (obstacle) de frontière Γ et soit Ω le domaine non borné complémentaire de $\bar{\mathcal{O}}$. On considère l'écoulement stationnaire lent (c'est à dire en négligeant les termes d'inertie) d'un fluide incompressible visqueux qui adhère à la paroi et qui tend à l'infini vers un écoulement uniforme donné $\underline{V}_\infty$. Les champs de vitesse et de pression $\underline{u}, p$ doivent satisfaire à :

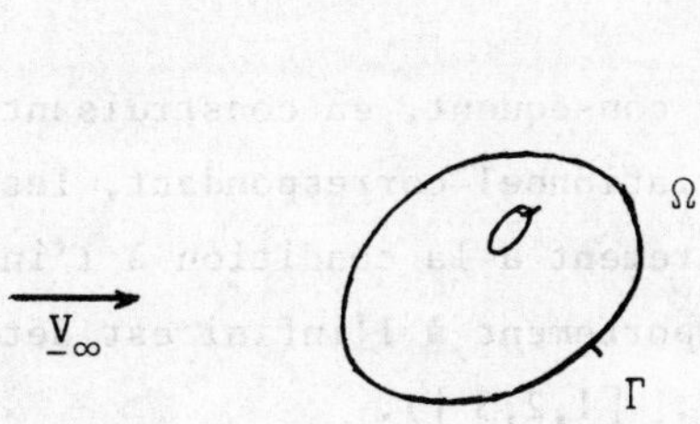

Figure 1

$$(1.1)\quad \begin{cases} 0 = -\dfrac{\partial p}{\partial x_i} + \Delta u_i & \text{dans } \Omega, i = 1,2 \\ \operatorname{div} \underline{u} = 0 & \text{"} \quad \text{"} \\ \underline{u} = 0 & \text{sur } \Gamma \\ \underline{u} \to \underline{V}_\infty \quad \text{si} & |x| \to \infty . \end{cases}$$

Le <u>paradoxe de Stokes</u> dit que le problème (1.1) n'a pas de solution sauf dans le cas trivial $\underline{V_\infty} = 0$. Autrement dit, avec les équations (1.1) on ne peu satisfaire simultanément à la condition aux limites de Dirichlet sur Γ et à la condition à l'infini. L'explication de ce phénomène est bien connue : du point de vue mathématique cela tient aux propriétés de la capacité d'un domaine extérieur en dimension 2 et à la complétion de $\mathcal{D}(\Omega)$ par rapport à l'intégrale de Dirichlet

$$(1.2)\qquad \|u\|^2 = \int_\Omega |\operatorname{grad} u|^2 \, dx .$$

On peut construire des suites de Cauchy de fonctions de $\mathcal{D}(\Omega)$ convergeant vers une fonction u non nulle à l'infini, (Fig. 2).

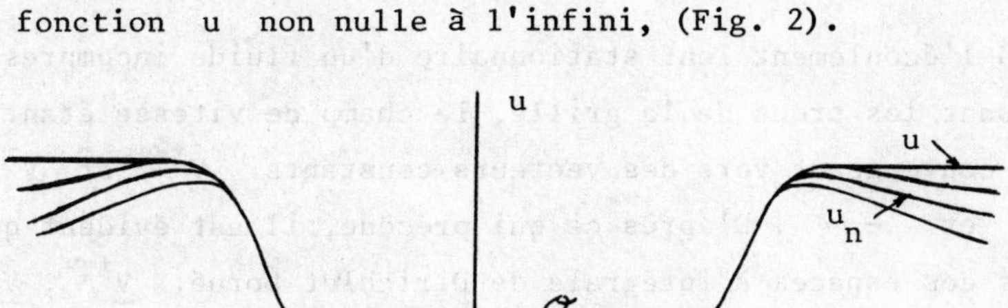

Figure 2

Par conséquent, en construisant un espace de Hilbert pour traiter le problème variationnel correspondant, les fonctions de l'espace ne satisfont pas nécessairement à la condition à l'infini, qui sera, en général violée, et le comportement à l'infini est déterminé par les autres données du problème (cf. [1,2,3]).

Du point de vue mécanique, l'écoulement lent existe bien pour des nombres de Reynols petits mais non nuls, les termes d'inertie négligés dans (1.1) jouent un rôle fondamental dans le comportement à l'infini. Les problèmes du type (1.1) apparaissent souvent pour l'obtention de certains termes de développements asymptotiques (cf. [4]).

2 - ECOULEMENT DE STOKES PERIODIQUE A TRAVERS UNE GRILLE

Dans l'espace R^3 des variables y_1, y_2, y_3 (ou R^N ; $N > 1$, ceci n'a pas d'importance), on considère une période de base rectangulaire ω, contenue dans le plan R^2 des variables y_1, y_2 et une paroi de frontière Γ entourant le plan $y_3 = 0$ percé de trous disposés ω-périodiquement. L'extérieur de la paroi se considère rempli de fluide. On désigne par G la partie de $\omega \times \mathbb{R}$ contenue dans le domaine fluide (voir figures 3,4, et l'on dira indifféremment ω-périodique ou G-périodique).

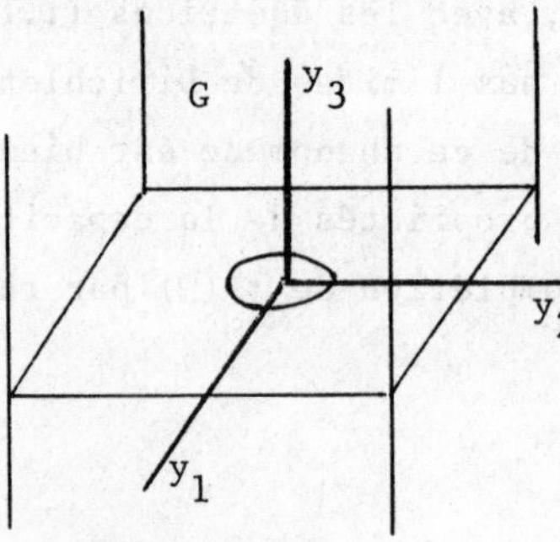

Figure 3

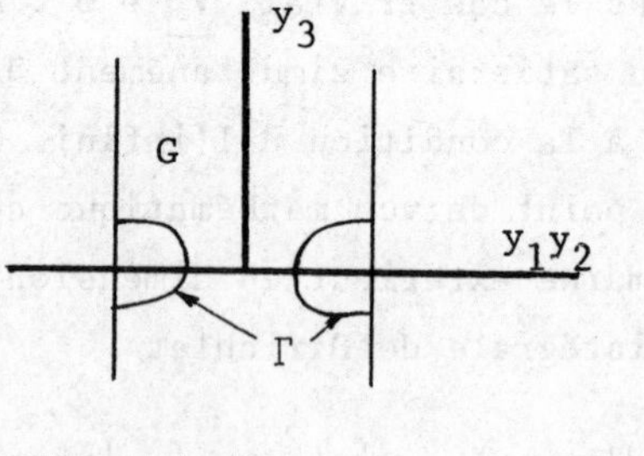

Figure 4

On s'intéresse à l'écoulement lent stationnaire d'un fluide incompressible visqueux traversant les trous de la grille, le champ de vitesse étant G-périodique et convergeant vers des vecteurs constants $\underline{V}^{+\infty}$ et $\underline{V}^{-\infty}$ pour $y_3 \to +\infty$ et $-\infty$. D'après ce qui précède, il est évident que en travaillant dans des espaces à intégrale de Dirichlet borné, $\underline{V}^{+\infty}$, $\underline{V}^{-\infty}$ ne pourront pas être en général des constantes données. Nous devons avancer un peu avant de poser des problèmes cohérents. De l'équation (1.1) on voit que si $\underline{u}$ est G-périodique, $\underline{\text{grad}}$ p l'est aussi ; ceci n'implique pas que p soit G-périodique, mais que p subit des accroissements bien déterminés en se déplaçant d'une arête de période dans les directions y_1, y_2. Mais si la convergence de $\underline{v}$ vers des constantes à l'infini est suffisamment rapide et régulière (nous verrons qu'elle l'est) ont voit que, à l'infini le $\underline{\text{grad}}$ p est nul, et donc p doit être G-périodique. Nous avons donc pour l'instant :

(2.1) $\quad 0 = \dfrac{-\partial p}{\partial y_i} + \Delta u_i \quad$ dans G

(2.2) $\quad \text{div}\, \underline{u} = 0 \quad$ " "

(2.3) $\quad \underline{u} = 0 \quad$ sur Γ

(2.4) $\quad u_i, p \quad$ sont G-périodiques.

3 - LEMMES SUR LE COMPORTEMENT A L'INFINI

Considérons des solutions-distributions de (2.1) - (2.4). Elles sont d'ailleurs des fonctions analytiques à l'intérieur des domaines ; la périodicité s'exprime en disant que pour y_3 fixé ce sont des fonctions définies sur le tore ω. En prenant la divergence de (2.1), compte tenu de (2.2), on a :

(3.1) $\quad -\Delta p = 0$.

L'étude à l'infini des solutions G-périodiques de l'équation (3.1) peut se faire en utilisant un lemme dû à TARTAR (cf. [5]) sur des solutions à gradient exponentiellement décroissant, ou les travaux de LANDIS, LAHIUROV et PANASENKO [6,7] utilisant notamment des adaptations du principe maximum. Nous utiliserons ici des tenchniques basées sur la transformée de Fourier par rapport à la variable y_3. On considère, avec des notations évidentes, l'espace de distributions tempérées $\mathscr{S}'(-\infty, +\infty, L^2(\omega))$ où ω désigne le tore et $\mathscr{S}(-\infty, +\infty, L^2(\omega))$ des fonctions C^∞ à valeurs dans $L^2(\omega)$ rapidement décroissantes, ainsi que toutes leurs dérivées, pour $y_3 \to \infty$.

LEMME 3.1

Soit $v(y_1, y_2, y_3)$ une fonction ω-périodique pour $y_3 > a$, où elle coïncide avec une distribution de $\mathcal{S}'(-\infty, +\infty, L^2(\omega))$ et satisfait à :

(3.2) $\qquad - \Delta v = f$

f coïncidant pour $y_3 > a$ avec une fonction de $\mathcal{S}(-\infty, +\infty, L^2(\omega))$.

Alors, il existe des constantes α, β telles que v s'écrit sous la forme

(3.3) $\qquad v = \alpha y_3 + \beta + v^{res}$

où v^{res} coïncide pour $y_3 > a$ avec une fonction de $\mathcal{S}(-\infty, +\infty, L^2(\omega))$.

Eléments de la démonstration

En prolongeant pour $y_3 < 0$, on passe de v, f à $\tilde{v}, \tilde{f}$ avec

(3.4) $$\left(- \frac{\partial^2}{\partial y_3^2} - \Delta_\omega \right) \tilde{v} = \tilde{f}$$

où $-\Delta_\omega$ est l'opérateur de Laplace par rapport aux variables y_1, y_2, dans le tore ω (i.e. ω-périodique). C'est un opérateur anticompact, autoadjoint positif mais non défini positif (0 est valeur propre, correspondant aux fonctions propres constantes). Il s'ensuit que $(\zeta - \Delta_\omega)^{-1}$ est holomorphe à valeurs dans $\mathcal{L}(L^2(\omega))$ pour $-\zeta$ différent des valeurs propres

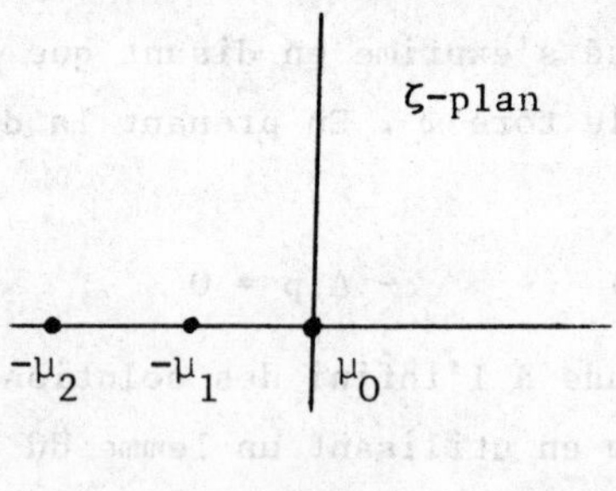

Figure 5

$$0 = \mu_o < \mu_1 \leqslant \mu_2 \leqslant \mu_3 \leqslant \ldots$$

et comme $-\Delta_\omega$ est autoadjoint, les pôles sont du premier ordre si bien que $\zeta(\zeta - \Delta_\omega)^{-1}$ n'a pas de pôle $\zeta = 0$ et est donc holomorphe dans un voisinage du demi axe réel positif. Alors $\lambda^2(\lambda^2 - \Delta_\omega)^{-1}$ est holomorphe pour λ réel. En faisant la transformée de Fourier $y_3 \to \lambda$ de (3.4) il vient

$$(\lambda^2 - \Delta_\omega)\, \hat{\tilde{v}} = \hat{\tilde{f}}$$

et en appliquant la fonction holomorphe construite on a

$$\lambda^2 \hat{\tilde{v}} \equiv \left(- \frac{\partial^2 \tilde{v}}{\partial y_3^2}\right)^{\wedge} = \lambda^2 (\lambda^2 - \Delta_\omega)^{-1} \hat{\tilde{f}}$$

et comme $\hat{\tilde{f}} \in \mathscr{S}$, le dernier membre est une fonction de classe C^∞ à valeurs dans $L^2(\omega)$; en plus, il appartient à $\mathscr{S}(-\infty,+\infty, L^2(\omega))$ comme on voit facilement du comportement pour λ grand de $(\lambda^2 - \Delta_\omega)^{-1}$ (noter que $-\Delta_\omega$ est un opérateur positif). Alors $(\partial^2 \tilde{v}/ \partial y_3^2) \in \mathscr{S}(-\infty, +\infty, L^2(\omega))$ et en intégrant deux fois on a (3.3) avec $\alpha, \beta \in L^2(\omega)$. On voit qu'elles sont des constantes en remplaçant (3.3) dans (3.2), en décomposant dans la base formée par les vecteurs propres et en faisant $y_3 \to +\infty$; les coefficients correspondants aux valeurs propres non nulles tendent vers 0.

LEMME 3.2

Soit $\underline{u},p$ une solution ω-périodique du système de Stokes (2.1), (2.2), (2.4) définie pour $y_3 > a$. On admet aussi que $\underline{\text{grad}}\ \underline{u} \in L^2(a,+\infty, L^2(\omega))$. Alors, il existe des constantes $\underline{u}^\infty, p^\infty$ telles que

$$(3.5) \qquad \begin{cases} \underline{u} = \underline{u}^\infty + \underline{u}^{res} \\ p = p^\infty + p^{res} \end{cases}$$

où $\underline{u}^{res}, p^{res}$ coïncide pour $y_3 > a$ avec une fonction de $\mathscr{S}(-\infty,+\infty,L^2(\omega))$.

Eléments de démonstration

Admettons que $(3.5)_2$ ait été prouvé. Alors $(3.5)_1$ résulte d'appliquer le lemme 3.1 à chaque composante u_i de (2.1). Pour prouver $(3.5)_2$, en prenant la divergence de $(3.5)_2$ on obtient l'équation de Laplace (3.1) ; pour lui appliquer le lemme 3.1 il faut s'assurer que p a une croissance tempérée à l'infini. Comme de l'hypothèse que $\underline{\text{grad}}\ \underline{u}$ est de carré sommable on voit que $\underline{\text{grad}}\ \underline{u}$ est une distribution tempérée, de (2.1), après prolongement on a:

$$\underline{\text{grad}}\ p \in \mathscr{S}'(-\infty,+\infty;H^{-1}(\omega))$$

d'où l'on déduit que $p \in \mathscr{S}'(-\infty,+\infty,L^2(\omega))$. Du lemme 3.1, on a :

$$p = \alpha y_3 + \beta + p^{res}$$

ensuite on établit que $\alpha = 0$ car $\alpha \neq 0$ est incompatible avec $\text{grad}\ \underline{u} \in L^2(a,\infty, L^2(\omega))$; on trouve une contradiction en intégrant la composante 3 de (2.1) dans $]a,b[\times \omega$ et en faisant $b \to \infty$. ■

4. DEUX PROBLEMES AUX LIMITES BIEN POSES

En revenant au système (2.1) - (2.4), sous l'hypothèse que la vitesse $\underline{u}$ ait ine intégrale de Dirichlet bornée nous connaissons la régularité des solutions ; la régularité à distance finie et classique, la régularité à l'infini étant donnée par le lemme 3.2 . Cela nous permet d'intégrer par parties pour définir des problèmes aux limites bien posés (la pression étant d'ailleurs toujours définie à une constante additive près).

Le caractère incompressible de l'écoulement (équation (2.2)) a des conséquences importantes : le flux à l'infini, a même valeur pour y_3 tendant vers $+\infty$ où $-\infty$ et c'est d'ailleurs égal au flux $\Phi(\underline{u})$ à travers le trou de la paroi (le flux latéral s'annule par périodicité) :

$$(4.1) \qquad |\omega| \, u_3^{+\infty} = |\omega| \, u_3^{-\infty} = \Phi(u) .$$

Construisons maintenant un espace de fonctions $\underline{u}$ de divergence nulle à intégrale de Dirichlet bornée. Ceci peut se faire de deux façons donnant deux espaces <u>distincts</u> (voir [8] pour d'autres situations analogues d'écoulements incompressibles dans des domaines avec des trous). On considère les deux espaces (non complets) ,

$$\dot{\mathscr{V}} = \{ \underline{v} \; ; \; C^\infty , \text{ G-pér., div } \underline{v} = 0, \, \underline{v}|_\Gamma = 0, \text{ avec } \underline{v} = 0 \text{ dans des voisinages de } y_j = \pm \infty \}$$

$$\dot{\mathscr{W}} = \{ \underline{v} \; ; \; C^\infty , \text{ G-pér., div } \underline{v} = 0, \underline{v}|_\Gamma = 0, \text{ avec } \underline{v} = \underline{cte} \text{ dans des voisinages de } y_3 = \pm \infty \}$$

ou l'on remarque par ailleurs que si $\underline{v} \in \dot{\mathscr{W}}$ les valeurs à l'infini de $\underline{v}$ satisfont à l'égalité de flux (4.1). Les espaces de Hilbert, $\mathscr{V}, \mathscr{W}$ sont les complétés de $\dot{\mathscr{V}}$ et $\dot{\mathscr{W}}$ pour la norme de Dirichlet :

$$\|\underline{v}\|^2 = \int_G \sum_{ik} \left\| \frac{\partial v_i}{\partial y_k} \right\|_{L^2(G)} .$$

Par ailleurs, en prenant la restriction à un domaine avec $|y_3|$ borné, on voit que si $\underline{v}$ appartient à $\mathscr{V}$ ou a $\mathscr{W}$, la restriction est de classe H^1 , si bien que les théorèmes de trace usuels sont valables. On voit alors que les vecteurs de $\mathscr{V}$ ont un flux nul à travers le trou ,

$$v \in \mathscr{V} \Longrightarrow \Phi(\underline{v}) = 0 = \int_{G \cap \{y_3 = o\}} v_3 \, dy_1, dy_2$$

condition qui n'est pas satisfaite en général par les $v \in \mathscr{W}$. Une étude plus fine montre que $\mathscr{V}$ est un sous-espace fermé de $\mathscr{W}$ de codimension 1 .

En revenant au problème (2.1) - (2.4), compte tenu du lemme 3.2, on voit que l'espace $\mathscr{V}$ sera approprié pour définir une formulation variationnelle d'un problème où la donnée sera le flux Φ à travers le trou (le problème aux limites sera non homogène). D'autre part, en intégrant par partie les termes de pression, on a :

$$\int_G \frac{\partial p}{\partial y_i} w_i \, dy = \int_G \frac{\partial}{\partial y_i} (p \, w_i) \, dy = \int_{\partial G} p \, w_i n_i \, dS$$

où $\underline{n}$ est la normale unitaire extérieure à G (ou a un autre domaine). On voit alors que, si l'on prend une fonction test $\underline{w} \in \mathscr{W}$, le terme de pression ne disparait pas dans la formulation variationnelle : il reste $(p^{+\infty} - p^{-\infty}) \, \Phi(\underline{w})$. C'est pourquoi $\mathscr{W}$ est un espace approprié pour définir une formulation variationnelle du problème dont la donnée soit la différence de pression entre $+\infty$ et $-\infty$. On arrive ainsi à définir les deux problèmes normalisés suivants (dont les solutions existent et sont uniques, la pression étant définie à une constante additive près :

<u>Problème en flux</u> : trouver $\underline{U}$,P G-périodiques, satisfaisant à (2.1) - (2.4) et

$$U_3 \longrightarrow 1 \qquad \text{si} \quad |y_3| \longrightarrow \infty \tag{4.4}$$

où seule la composante 3 de la vitesse est donnée à l'infini (4.4). Le champ de vitesse et pression est alors complètement déterminé ; en particulier si $\underline{a}$ est un champ vectoriel G-périodique, de divergence nulle et constant pour $|y_3|$ grand, satisfaisant à (4.4) ce problème équivaut à

trouver $\underline{U} = \underline{a} + \underline{V}$, $\underline{V} \in \mathscr{V}$ tel que

$$\int_G \frac{\partial U_i}{\partial y_i} \frac{\partial v_i}{\partial y_i} \, dy = - \int_G \frac{\partial a_i}{\partial y_i} \frac{\partial v_i}{\partial y_i} \, dy \quad \forall \, \underline{v} \in \mathscr{V}.$$

Nous voyons donc que la composante 3 peut être donnée (il n'y a pas de paradoxe de Stokes pour elle) mais les quatre constantes $U_1^{+\infty}$, $U_2^{+\infty}$, $U_1^{-\infty}$, $U_2^{-\infty}$ sont alors déterminées de façon unique : il y a paradoxe de Stokes pour elles. D'autre part, nous avons :

Problème en différence de pression : trouver $\underline{U},P$ satisfaisant à (2.1)-(2.4) et à :

$$P^{+\infty} - P^{-\infty} = 1 \ .$$

Ce problème est équivalent à :

trouver $U \in \mathscr{W}$ tel que

$$\text{(4.6)} \qquad \int_G \frac{\partial U_i}{\partial y_j} \frac{\partial v_i}{\partial y_j} \, dy = - \Phi(v) \qquad \forall \underline{v} \in \mathscr{W}$$

où le second membre désigne naturellement le flux de la fonction test à travers le trou ; c'est une fonctionnelle linéaire bornée sur $\mathscr{W}$ grâce au théorème de trace.

Le problème en différence de pression ressemble en fait à la loi de Darcy pour les écoulements fluides dans des milieux poreux ; la différence de pression (au lieu du gradient moyen) détermine l'écoulement périodique.

Par ailleurs, on voit immédiatement en prenant $\underline{v} = \underline{U}$ dans (4.6) que si $\underline{U}$ est la solution du problème en différence de pression, il existe une constante $\gamma > 0$ définie par la géométrie telle que :

$$\Phi(\underline{U}) = - \gamma$$

si bien que la solution du problème en flux est $-1/\gamma$ fois la solution du problème en différence de pression. Nous voyons que les deux études avec les espaces $\mathscr{V}$ ou $\mathscr{W}$ se rejoignent finalement.

5 - APPLICATION. PROBLEME ASYMPTOTIQUE A TRAVERS UNE GRILLE FINE

On considère un domaine Ω traversé par la grille Σ autour du plan $x_3 = 0$ (fig. 6). Cette grille est homothétique de rapport ε de celle des sections précédentes. On se donne un champ de vitesse $\underline{b}$ sur $\partial\Omega$ (mis à part la grille) tel que le flux total qui traverse la grille, compte renu de l'imcompressibilité soit $F \neq 0$:

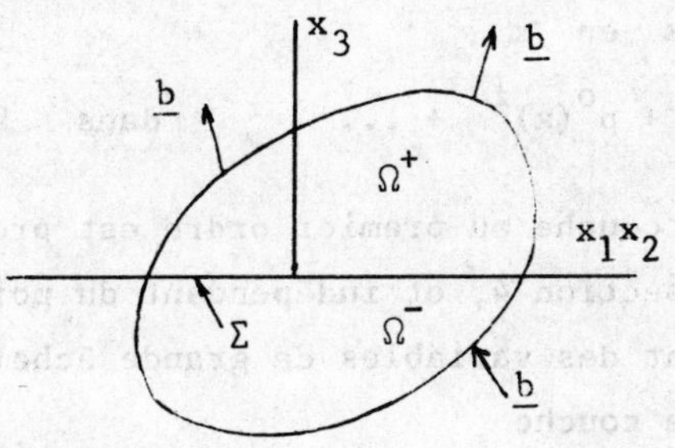

Figure 6

$$(5.1) \qquad \int_{\partial\Omega\cap\{x_3>o\}} \underline{b}.\underline{n}\, ds = - \int_{\partial\Omega\cap\{x_3<o\}} \underline{b}.\underline{n}\, ds = F \neq 0 \ .$$

On considère alors, pour ε petit le problème de Stokes : trouver $\underline{u}^\varepsilon$, p^ε satisfaisant à :

$$(5.2) \qquad 0 = - \frac{\partial p^\varepsilon}{\partial x_i} + \Delta u_i^\varepsilon$$

$$(5.3) \qquad \text{div } \underline{u}^\varepsilon = 0$$

$$(5.4) \qquad \underline{u}^\varepsilon = 0 \qquad \text{sur } \partial\Omega$$

$$(5.5) \qquad u^\varepsilon = 0 \qquad \text{sur la } \varepsilon\text{-grille} \ .$$

On cherche des développements asymptotiques formels dans les régions Ω^+, Ω^- et dans la couche au voisinage de Σ avec les hypothèses habituelles de $\varepsilon\omega$-périodicité locale de l'homogénéisation. Comme $F \neq 0$, le flux par unité de surface à travers Σ est d'ordre $O(1)$ pour ε petit. En faisant le changement $y = x/\varepsilon$ au voisinage de la grille, il vient :

$$(5.6) \qquad \frac{\partial}{\partial x_i} = \frac{1}{\varepsilon} \frac{\partial}{y_i} \quad ; \quad \Delta_x = \frac{1}{\varepsilon^2} \Delta_y$$

et comme le problème en y est à divergence nulle, il doit avoir un terme de pression (multiplicateur de Lagrange) ; (5.6) montre alors que la différence de pression à la traversée de la couche de la grille doit être $O(\varepsilon^{-1})$, si bien que le développement de la pression dans Ω^+ et Ω^- doit commencer par des termes en ε^{-1}. Comme d'autre part les champs de vitesse et pression dans Ω^+ et Ω^- sont naturellement $O(1)$ (la pression à une constante additive près), on voit que les termes principaux de pression dans $\Omega^{\pm}$ sont constants en x

$$(5.7) \qquad p^{\varepsilon} = C^{\pm} \frac{1}{\varepsilon} + p^{o}(x)^{\pm} + \ldots \qquad \text{dans} \quad \Omega^{\pm}$$

et l'écoulement dans la couche au premier ordre est proportionnel à la solution du problème de la section 4, et indépendant du point sur la grille (c'est à dire indépendant des variables de grande échelle). La condition de flux donne alors dans la couche

$$(5.8) \qquad \underline{u}^{\varepsilon}(x) = \frac{F}{|\Sigma|} \underline{U}(x/\varepsilon) + O(\varepsilon) \qquad \text{au voisinage de } \Sigma .$$

Alors les champs de vitesse à l'ordre $O(1)$ dans $\Omega^{\pm}$ sont donnés par la solution $\underline{u}^{o}, p^{o}$ de

$$0 = -\frac{\partial p^{o}}{\partial x_i} + \Delta u_i \qquad \text{dans} \quad \Omega^{\pm}$$

$$0 = \operatorname{div} \underline{u}^{o} \qquad " \qquad "$$

$$\underline{u}^{o} = 0 \qquad \text{dans} \qquad \partial\Omega \cap \{x_3 > 0\}$$

$$\underline{u}^{o} = \frac{F}{|\Sigma|} \underline{U}^{+\infty} \qquad \text{dans} \qquad x_3 = 0^{+}$$

et un problème analogue dans Ω^{-}.

6 - REMARQUES FINALES

Le fait que l'écoulement asymptotique soit indépendant des variables de grande échelle peut se déduire facilement par des considérations énergétiques. L'énergie dissipée par viscosité est d'ordre ε^{-1} et se produit asymptotiquement au voisinage de la grille. On peut voir que mis à part le facteur ε^{-1}, elle est proportionnelle à :

$$\int_{\Sigma} \varphi(x_1, x_2)^2 \, dx_1 \, dx_2$$

où φ désigne le flux par unité de surface à travers Σ. Comme $F = \int_{\Sigma} \varphi(x_1, x_2) \, dx_1 \, dx_2$ est donnée, la minimisation de l'énergie dissipée donne φ indépendant de x_1, x_2.

Le problème de la convergence vers la solution asymptotique formelle est ouvert. Nous ingorons s'il entre dans le cadre de la Γ-convergence [9].

D'autres problèmes ouverts sont les problèmes ω-périodiques non linéaires soit en conservant les termes d'inertie, soit pour des fluides viscoplastiques de Bingham (qui semblent être intéressants car doivent donner lieu à des "lois du type Darcy non linéaires à seuil"). La méthode de la transformée de Fourier ne me semble pas être applicable à des problèmes non linéaires. Mais au point de vue application, il faut noter que si le problème dans les variables macroscopiques x est non linéaire, à nombre de Reynolds fixé indépendant de ε, le problème local en y est linéaire, exactement celui qui a été étudié, et le problème asymptotique est analogue à celui du cas linéaire.

Le problème de la section 5 m'a été proposé par F. MURAT. Une version modifiée se trouve dans [10,11] et d'autres problèmes d'homogénéisation de frontières peuvent être consultés dans [12,13].

BIBLIOGRAPHIE

[1] J. DENY et J.L. LIONS : Les espaces du type Beppo Levi, Ann. Inst. Fourier 5 (1953-4) p. 305-370.

[2] L. HORMANDER et J.L. LIONS : Sur la complétion par rapport à une intégrale de Dirichlet, Math. Scand., 4 (1956), p. 259-270.

[3] O.A. LADYZHENSKAYA : Mathematical Theory of viscous incompressible flow, Gordon an Breach, (New York) 1963.

[4] M. VAN DYKE : Perturbation methods in fluid mechanics, Academic Press, New York (1964).

[5] J.L. LIONS : Some Methods in the Mathematical Analysis of Systems and their Control, Gordon and Breach, New York, (1981).

[6] E.M. LANDIS et S.S. LAHTUROV : On the behaviour at infinity of solutions of elliptic equations that are periodic in all variables except one. Sov. Math. Doklady 21 (1980), p. 211-214 (= Dokl. Akad. Nauk 250, 1980, n° 4).

[7] E.M. LANDIS et G.P. PANASENKO : A theorem on the asymptotics of solutions of elliptic equations with coefficients periodic in all variables except one. Soviet Math. Dokl., 18 (1977), p. p. 1140-1142 (= Dokl. Akad. Nauk, 235, 1977, n° 6.

[8] J.G. HEYWOOD : On uniqueness questions in the theory of viscous flows. Acta Math. 136 (1976), p. 61-102.

[9] E. DE GIORGI : Convergence problems for functionals and operators, in Recent Methods in Nonlinear-Analysis, ed. De Gogi, Magenes, Mosco, Pitagora, (Rome) 1979, p. 131-188.

[10] E. SANCHEZ-PALENCIA : Un problème d'écoulement lent d'un fluide visqueux incompressible à travers une paroi finement perforée. Ecole d'Analyse Numérique E.D.F., C.E.A., I.N.R.I.A. sur l'homogénéisation, 1983 (à paraître, Ed. Eyrolles).

[11] E. SANCHEZ-PALENCIA : Ecoulement lent d'un fluide visqueux incompressible à travers une paroi perforée périodiquement. Compt. Rend. Acad. Sci., série II, 296, (1983), p. 1629-1632.

[12] E. SANCHEZ-PALENCIA : Non homogeneous media and vibration theory. Springer, Berlin (1980).

[13] J. SANCHEZ-HUBERT et E. SANCHEZ- PALENCIA : Acoustic Fluid Flow through holes and Permeability of Perforated Walls. Jour. Math. Anal. Appl., 87 (1982), p. 427-453.

E. SANCHEZ-PALENCIA
LABORATOIRE DE MECANIQUE THEORIQUE
Université Pierre et Marie Curie
Tour 66 - 4, Place Jussieu
75230 - PARIS CEDEX 5

C SBORDONE

Rearrangement of functions and reverse Hölder inequalities

1 - INTRODUCTION

The Hardy-Littlewood maximal function of a non-negative locally integrable function f on R^n is given by

$$Mf(x) = \sup_{x\in Q} \frac{1}{|Q|} \int_Q f(y)dy ,$$

where the supremum is taken over all open cubes Q containing x with sides parallel to the axes.

In 1968 C. HERZ [10] proved that there exist two constants $c_1 = c_1(n)$, $c_2 = c_2(n)$ such that for any $t > 0$

$$c_1 (Mf)^*(t) \leqslant \frac{1}{t} \int_o^t f^*(s)ds \leqslant c_2 (Mf^*(t)$$

where $g^*(t)$ is the non-increasing rearrangement of the measurable function $g : R^n \rightarrow [0, +\infty[$ defined by

$$g^*(t) = \sup_{|E| = t} \inf_E g .$$

This result has been extended in [18] to the case where the Lebesgue measure is replaced by a "doubling" measure wdx, $w \in L^1_{loc}(R^n)$, $w \geqslant 0$ (see sec. 2).

These results can be applied to deduce a higher integrability theorem from a weighted reverse Hölder inequality of the type $(q > 1)$

$$(1.1) \qquad \frac{1}{w(Q)} \int_Q f^q wdx \leqslant B(\frac{1}{w(Q)} \int_Q fw\, dx)^q \quad \forall Q \text{ cube}$$

which generalizes some theorems of MUCKENHOUPT [14], COIFMAN-FEFFERMAN [3], GEHRING [7], (theo. 4.1).

(*) This work has been performed under a national research program of Italian M.P.I. (40% - 1982).

In particular, we reduce (1.1) to a reverse Hölder inequality on f^*, the non-increasing rearrangement of f with respect to the measure $w dx$.

In one dimension the reverse Hölder inequality for f^* is treated in [16] via Hardy's inequality, which gives the higher integrability theorem for f^* and f.

In sec 5, following an idea of L. TARTAR, we use a weighted Hardy's inequality to get a "sharp" result, giving the optimal integrability exponent of f^*.

Reverse integral inequalities with different support have also recently been investigated. See the references in [18].

2 - PRELIMINARIES AND NOTATION

We state the definitions, notation and results we will use in the following.

All cubes will have sides parallel to the coordinate axes and δQ will denote the cube with the same center as Q but δ-times its size.

Let Q_o be a cube in R^n : by $|Q_o|$ we mean its Lebesgue measure.

Let $w \in L^1(Q_o)$ be a non-negative function and set for $E \subset Q_o$

$$w(E) = \int_E w dx .$$

For any measurable non-negative function f, define for $y > 0$

$$\lambda_f(y) = w(\{x \in Q_o : f(x) > y\})$$

the distribution function of f with respect to the measure $w dx$ on Q_o.

The non-increasing rearrangement of f with respect to $w dx$ is defined for $t \in (0, w(Q_o))$ by

$$f^*(t) \inf \{ y > 0 : \lambda_f(y) \leqslant t \} .$$

The map $f \longrightarrow f^*$ is an example of G-operator in the sense of De GIORGI [4], since it can be shown that [13]

$$f^*(t) = \sup_{\substack{E \subset Q_o \\ w(E) = t}} \inf_E f .$$

Let us now list the basic properties of the non-increasing rearrangement which will be useful in the sequel.

The following relations hold [2] :

(2.1) $\int_0^{w(Q_o)} f^*(s)^p ds = \int_{Q_o} f^p w dx \qquad \forall p \geq 1$

(2.2) for $t \leq w(Q_o)$

$$\int_0^t f^*(s)ds = \sup_{\substack{E \subset Q \\ w(E) \leq t}} \int_E fw\,dx$$

(2.3) for any $\varphi : R^+ \longrightarrow R^+$ continuous and non-decreasing

$$(\varphi \circ f)^* = \varphi \circ f^* .$$

(2.4) Set $E(t) = \{x \in Q_o : f(x) > t\}$; then

$$\int_{E(t)} (f-t)dx = \int_t^\infty \lambda_f(\tau)d\tau = \int_0^{\lambda_f(t)} f^*(\sigma)d\sigma - t\lambda_f(t).$$

There is an abundant literature about the notion of rearrangement and its applications which goes back to [9].

For a recent result which shows its relations with the study of weak convergence in L^p spaces see [12].

The following definition will be important :

(2.5) let $w \in L^1(Q_o)$, $w \geq 0$; then $w dx$ is a <u>doubling measure</u> if there exists $d > 0$, such that $w(2Q) \leq d\, w(Q), \forall Q$ cube $\subset \frac{1}{2} Q$.

<u>Remark 2.1</u>

If $w \in A_p$, then ([15]) it is easy to show that $w dx$ is doubling.

3 - <u>REARRANGEMENT OF THE MAXIMAL FUNCTION</u>

The maximal function, introduced by Hardy and Littlewood [8], is an important tool in Harmonic Analysis and related fields.

Let Q_o be a cube in R^n and $w \in L^1(Q_o)$ a non-negative weight such that the doubling condition (2.5) is satisfied.

For $f \in L^1(Q_o, w dx)$, $f \geq 0$, let us define the weighted local maximal function of f relative to Q_o as

(3.1) $$Mf(x) = \sup_{x \in Q \subset Q_o} \frac{1}{w(Q)} \int_Q fw\,dx , \quad x \in Q_o ,$$

where the supremum is taken over all cubes $Q \subset Q_0$, such that $x \in Q$.

In this section we are interested in the relations between $(Mf)^*$, the non-increasing rearrangement of Mf with respect to wdx on Q_0, and the average f^{**} of f^*:

$$f^{**}(t) = \frac{1}{t} \int_0^t f^*(s)\, ds.$$

In the case $w = 1$, the fact that $(Mf)^*$ is dominated by f^{**} goes back to Hardy and Littlewood [8], while it was only in 1968 that Herz [10] showed that the converse is also true:

<u>THEOREM 3.1</u> - (Herz)

<u>Let</u> $w(x) = 1$, $f \in L^1(Q_0)$, $f > 0$ <u>then</u> $\forall t < |Q_0|$

$$3^{-n}(Mf)^*(t) \leqslant f^{**}(t) \leqslant (2^n + 1)\ (Mf)^*(t)$$

<u>where</u> M <u>is the classical local maximal operator</u>.

A proof of this theorem has been given recently in [1].

In the following we shall give a weighted version of Herz's theorem, the main tools for its proof being:

1) a Calderon-Zygmund <u>decomposition</u> of the cube Q_0 in terms of the averages of f with respect to $w(x)\,dx$ [6],

2) a Besicovitch-De Giorgi <u>covering</u> lemma, implying a weak type inequality for Mf. [5], [6].

<u>LEMMA 3.1</u> (Calderon-Zygmund)

<u>Let</u> $w \in L^1(Q_0)$, $w > 0$ <u>satisfy</u> (2.5) <u>and let</u> $f \in L^1(Q_0, wdx)$. <u>Then, for any</u> $\lambda > 0$ <u>there exists a collection</u> $(Q_j)_{j \in N}$ <u>of disjoint cubes</u> $Q_j \subset Q_0$, <u>such that</u>

$$\lambda < \frac{1}{w(Q_j)} \int_{Q_j} fx\, dx \leqslant d\, \lambda$$

(3.2)

$$f(x) \leqslant \lambda \quad \text{a.e. in } Q_0 \setminus \bigcup_j Q_j .$$

<u>LEMMA 3.2</u> (Besicovitch - De Giorgi)

<u>Let</u> A <u>be bounded in</u> R^n. <u>For each</u> $x \in A$ <u>a cube</u> Q_x <u>centered at</u> x <u>is given. Then there exists a sequence</u> $(Q_j)_{j \in N}$ <u>contained in</u> $(Q_x)_{x \in A}$ <u>such</u>

that

$$A \subset \bigcup_j Q_j$$

$$\sum_j \chi_{Q_j}(y) \leqslant \theta \qquad \forall y \in R^n ,$$

where θ depends only on the dimension n .

REMARK 3.1

An immediate consequence of lemma 3.2 is the following weak type inequality:

$$(3.4) \qquad w(\{Mg > \lambda\}) \leqslant \frac{c_o}{\lambda} \int_{Q_o} gwdx$$

for $g \in L^1(Q_o, wdx)$ and $g \geqslant 0$, where c_o depends only on the constant d in (2.5) and the dimension n .

We can now state the following generalization of Herz's theorem 3.1

THEOREM 3.2

Let w be non-negative and satisfy (2.5). Let $f \in L^1(Q_o wdx)$, $f \geqslant 0$. Then, for any $t < w(Q_o)$

$$c_1(Mf)^*(t) \leqslant f^{**}(t) \leqslant c_2(Mf)^*(t),$$

where c_i are positive constants depending only on d and n .

PROOF ([18]).

4 - WEIGHTED REVERSE HÖLDER INEQUALITIES

Let Q_o be a cube in R^n with sides parallel to the axes. For $p > 1$ a non-negative weight function $w \in L^1(Q_o)$ belongs to the A_p class of Muckenhoupt over Q_o [14], if there exists $A > 1$ such that

$$(4.1) \qquad \frac{1}{|Q|} \int_Q wdx \left(\frac{1}{|Q|} \int_Q w^{-\frac{1}{p-1}} dx\right)^{p-1} \leqslant A$$

for any parallel cube $Q \subset Q_o$.

For $q > 1$ a non-negative function $f \in L^q(Q_o)$ belongs to the G_q class of Gehring over Q_o [7], if there exists $G > 1$ such that

$$(4.2) \qquad \frac{1}{|Q|} \int_Q f^q dx \leqslant G \left(\frac{1}{|Q|} \int_Q f dx\right)^q$$

for any parallel cube $Q \subset Q_o$.

Both (4.1) and (4.2) are "reverse" Hölder inequalities.

Our aim is to prove that the "backward propagation" of the A_p condition :

$$w \in A_p \Longrightarrow \exists \delta > 0 : w \in A_{p-\delta}$$

due to Muckenhoupt [14] and Coiffman-Fefferman [3], and the "forward propagation" of the G_q condition :

$$f \in G_q \Longrightarrow \exists \varepsilon > 0 : f \in G_{q+\varepsilon}$$

due to Gehring [7], are both corollaries of the following result : [18]

THEOREM 4.1

Let $w \in L^1(Q_o)$ *satisfy* (2.5), *and* $f \in L^q(wdx)$, $q > 1$, *satisfy*

$$(4.3) \qquad \frac{1}{w(Q)} \int_o f^q w\,dx \leq B\left(\frac{1}{w(Q)} \int_o fw\,dx\right)^q$$

for any cube $Q \subset Q_o$. Then, there exists $p > q$, $p = p(q,B,d,n)$ such that

$$(4.4) \qquad \frac{1}{w(Q)} \int_Q f^p w dx \leq C\left(\frac{1}{w(Q)} \int_Q f\,wdx\right)^p$$

for any cube $Q \subset Q_o$; where $C = C(p,q,B,d,n)$.

Remark 4.1

As a first corollary of theo. 4.1, choosing $w(x) = 1$ (and so $d = 2^n$), we obtain Gehring's theorem.

Remark 4.2

From the proof of theo. 4.1 it follows that if f verifies a reverse Hölder inequality of the type G_q, then its nonincreasing arrangement f^* verifies another reverse Hölder inequality [18], [16], [17].

Let us now assume that $w \in A_p$, $p > 1$. It is well known (see remark 2.1) that in this case the measure wdx is doubling, and we can prove

COROLLARY 4.1.

If $w \in A_p$, *then there exists* $\delta > 0$ *such that* $w \in A_{p-\delta}$.

Proof By assumption, for any cube $Q \subset Q_o$ 4.1 holds. Set $p' = p/(p-1)$ and so (4.1) becomes

$$\left(\frac{1}{|Q|}\int_Q w\,dx\right)^{p'-1}\left(\frac{1}{|Q|}\int_Q w^{1-p'}\right) \leqslant A^{p'-1}$$

that is

$$\frac{1}{w(Q)}\int_Q w^{-p'}\,w\,dx \leqslant A^{p'-1}\left(\frac{1}{w(Q)}\int_Q w^{-1}\,w\,dx\right)^{p'}$$

which is a weighted $G_{p'}$ condition for w^{-1} with respect to wdx.

Then by theo. 4.1 we deduce a $G_{p'+\varepsilon}$ condition for w^{-1} with respect to wdx, which, with an inverse procedure, imples $w \in A_{p-\delta}$ for a certain δ.

5 - A SHARP RESULT IN ONE DIMENSION

The aim of this section is to prove the following

THEOREM 5.1

Let $f : (0,m) \longrightarrow R^+$ be a non-increasing function verifying

$$(5.1)\qquad ⨍_a^b f^2 dx \leqslant B\left(⨍_a^b f dx\right)^2 \qquad \forall (a,b) \subseteq (0,m).$$

Set $\varepsilon(B) = \sqrt{\frac{B}{B-1}} - 1$. Then, for $p \in [2, 2+\varepsilon(B)[$ there exists $c_p = c_p(B) > 0$ such that

$$(5.2)\qquad ⨍_a^b f^p dx \leqslant c_p\left(⨍_a^b f dx\right)^2 \qquad \forall (a,b) \subseteq (0,m),$$

with $c_p(B) \longrightarrow +\infty$ as $p \longrightarrow 2+\varepsilon(B)$.

The result is sharp.

Let us begin with the following

LEMMA 5.1

Let $f : (0,m) \longrightarrow R^+$ verify (5.1) and $f \in L^\infty(0,m)$. Set

$$(5.3)\qquad \gamma(\beta) = 1 + \frac{4\beta B}{(1-\beta)^2} \quad \text{for } \beta > \beta_0 = 1 - 2B + 2\sqrt{B^2 - B},\ \beta < 0.$$

Then

$$\gamma(\beta)\ ⨍_a^b (t-a)^\beta f^2(t)\,dt \leqslant (b-a)^\beta ⨍_a^b f^2(t)\,dt \qquad \forall (a,b) \subseteq (0,m).$$

Proof

Set $\qquad g(x) = ⨍_a^x f(t)\,dt;$

then by the weighted Hardy's inequality ([11])

$$(5.4) \qquad \int_a^b (x-a)^\beta g^2(x)dx \leq \frac{4}{(1-\beta)^2} \int_a^b (x-a)^\beta f^2(x)dx$$

and so, using (5.1) and the Fubini theorem we have the bounds :

$$(5.5) \qquad \int_a^b (x-a)^\beta g^2 dx \geq \frac{1}{B} \int_a^b (x-a)^{\beta-1} \int_a^x f^2 dt\, dx =$$

$$= \frac{1}{B} \int_a^b f^2(t) \int_t^x (x-a)^{\beta-1} dx\, dt =$$

$$= \frac{1}{\beta B} \int_a^b f^2(t) [(b-a)^\beta - (t-a)^\beta] dt.$$

From (5.4) and (5.5), since $\beta < 0$, we deduce

$$\frac{4\beta B}{(1-\beta)^2} \int_a^b (t-a)^\beta f^2 dt \leq \int_a^b [(b-a)^\beta - (t-a)^\beta] f^2(t) dt$$

that is, (5.2).

Another result is the following [9].

LEMMA 5.2

Let $f : (0,m) \longrightarrow R^+$ be a non-increasing function. Then for $-1 < \beta < 0$, $(a,b) \subseteq (0,m)$:

$$\left(\int_a^b f^{2/(1+\beta)} dt\right)^{\beta+1} \leq \int_a^b (t-a)^\beta f^2(t)dt .$$

We are now able to give the

Proof (of theo. 5.1)

Let f verify the assumptions of theo. 5.1 . Then, by convolution, we can construct a sequence $f_h \in L^\infty(0,m)$ such that $f_h \longrightarrow f$ in $L^2(0,m)$ and f_h verify (5.1) with the same constant B .

By lemma 5.1 we deduce for $\beta \in]\beta_0, 0[$

$$(5.6) \qquad \gamma(\beta) \int_a^b (t-a)^\beta f_h^2(t)dt \leq (b-a)^\beta \int_a^b f_h^2(t)\, dt$$

where β and $\gamma(\beta)$ are given in 5.3 .

Passing to the limit as $h \longrightarrow \infty$ in (5.6) we deduce

$$\gamma(\beta) \int_a^b (t-a)^\beta f^2 dt \leq (b-a)^\beta \int_a^b f^2 dt$$

and also, by lemma (5.2)

$$\left(\int_a^b f^{2/(1+\beta)} dt \right)^{\beta+1} \leqslant \frac{(b-a)^\beta}{\gamma(\beta)} \int_a^b f^2 \, dt$$

for $\beta > \beta_0$, i.e. (5.2), with $p = 2/(1+\beta)$.

The function $f(t) = t^{-\alpha}$ with $\alpha = 1 - B + \sqrt{B^2 - B}$ verifies (5.1) and (5.2) and so our result is sharp.

REFERENCES

[1] C. BENNETT - R. SHARPLEY : Weak type inequalities for H^p and BMO, Proc. Symp. Pure math. XXXV (1979).

[2] H.M. CHUNG - R. HUNT - D.S. KURTZ : The Hardy-Littlewood maximal function on L(p.q) spaces with weights, Indiana Un. Math. J. 31, 1, (1982).

[3] R.R. COIFMAN - C. FEFFERMAN : Weighted norm inequalities for maximal functions and singular integrals, Studia Math. 51 (1974).

[4] E. DE GIORGI : Generalized limits in Calculus of Variations, Quaderno Sc. Norm. Sup. "Topics in Functional Analysis 1980, Pisa (1982).

[5] E. DE GIORGI - F. COLOMBINI - L.C. PICCININI : Frontiere orientate di misura minima e questioni collegate, Quaderno Sc. Norm. Sup. Pisa (1972).

[6] M. DE GUZMAN : Real variable methods in Fourier Analysis, North Holland Math. Studies 46 (1981).

[7] F.W. GEHRING : The L^p-integrability of the partial derivatives of a quasiconformal mapping, Acta Math. 130 (1973).

[8] G.H. HARDY - J.E. LITTLEWOOD : A maximal function theorem with function theoretical applications, Acta Math. 54 (1930).

[9] G.H. HARDY - J.E. LITTLEWOOD - G. POLYA : Inequalities Cambridge Univ. Press (1964).

[10] C. HERZ : The Hardy-Littlewood maximal theorem, Symp. on Harmonic Analysis, Univ. of Warwick (1968).

[11] A. KUFNER - H. TRIEBEL : Generalizations of Hardy's inequality, Conf. Sem. Mat. Un. Bari 156 (1978).

[12] L. MIGLIACCIO : Sur une condition de Hardy-Littlewood-Polya, C.R. Acad. Sc. Paris 297 (1983).

[13] G. MOSCARIELLO : Riordinamenti e operatori di tipo G , to appear

[14] B. MUCKENHOUPT : Weighted norm inequalities for the Hardy maximal function, Trans. Amer. Math. Soc. 165 (1972).

[15] B. MUCKENHOUPT - R.L. WHEEDEN : Weighted bounded mean oscillation and the Hilbert transform, Studia math. LIV (1976).

[16] C. SBORDONE : Higher integrability from reverse integral inequalities, Proc. Intern. Meeting - Methods of Functional Analysis and Theory of Elliptic Equations, Napoli, Liguori (1983).

[17] C. SBORDONE : Rearrangement of functions and reverse Jensen inequalities, Lecture Note Summer Institute Amer. Math Soc. Berkeley (1983).

[18] C. SBORDONE : Some reverse integral inequalities, Atti Academia Pontaniana, Napoli (to appear).

Carlo SBORDONE
Istituto Matematico R. Cacciopoli
Universita di Napoli
Via Mezzocannone 8
80134 - NAPOLI (Italy)

S SPAGNOLO

Global solvability in Banach scales of weakly hyperbolic abstract equations

§ - 1 INTRODUCTION

In this lecture we present a result, jointly obtained with A. AROSIO (see [1]), concerning the existence of a global solution for the initial value problem

(1) $u'' + A(t)\, u = 0 \qquad (o \leqslant t \leqslant T)$

(2) $u(0) = v_o \;,\; u'(0) = v_1 \;,$

where $A(t)$ is a family of non-negative self-adjoint operators in a Hilbert space H.

More precisely, we assume that

(3) $A \in L^1(0,T \;;\; \mathcal{L}(V,V'))$

(4) $< A(t)\, v,w > = < A(t)w,v >$

and

(5) $< A(t)v,v > \;\geqslant 0 \;,$

for some given reflexive Banach space V, continously and densely embedded in H. We denote by V' the antidual of V and by $< , >$ the duality map on $V' \times V$. Moreover we identify as usual H' with H, via the Riesz isomorphism, in order to get the Hilbert triplet

$$V \subseteq H \subseteq V' \,.$$

Under these assumptions we say that (1) is a weakly hyperbolic equation, whereas we say that Eq.(1) is strictly hyperbolic when conditions (3) and (5) are replaced respectively by :

(6) $A \in BV(0, T \;;\; \mathcal{L}(V,V'))$

and by

(7) $< A(t)v,v > \;\geqslant \nu |v|^2_V \quad$, for some $\nu > 0$.

When Eq.(1) is strictly hyperbolic it is not difficult to prove an "a priori" energy estimate, which leads to the well-posedness of Pb.{(1), (2)} in $V \times H$, in the sense that, for every $v_0 \in V$ and $v_1 \in H$, there exists a unique solution $u(t)$ in the space $C(0,T;V) \cap C^1(0,T;H) \cap W^{2,\infty}(0,T;V')$.

This conclusion may be false if one of the conditions (6) and (7) is not satisfied, as it is shown in [3] and [5] by means of two examples concerning a differential equation as

(8) $$u_{tt} - c(t)\, u_{xx} = 0 \quad \text{with} \quad c(t) \geqslant 0 .$$

In the first of these examples the function $c(t)$ is continuous and $\geqslant 1$, in the second one $c(t)$ is indefinitely differentiable and $\geqslant 0$: in both examples the Cauchy problem for Eq.(8) has no local solution in the space of distributions though the initial date are C^∞ functions.

One the other hand, the Cauchy problem for Eq.(8) is always well-posed in the space $\mathscr{A}(\mathbb{R}_x)$ of the real analytic functions (see [3] generally, the Cauchy problem for the equation

(9) $$u_{tt} - (c(x,t)\, u_x)_x = 0 \qquad (x \in \mathbb{R},\ t \in [0,T])$$

is well-posed in $\mathscr{A}(\mathbb{R}_x)$ whenever $c(x,t)$ is a non-negative function, integrable in t for every x and real analytic in x uniformly with respect to t (see [7] and [4]).

Hence we are induced to ask whether in the general abstract case of Eq.(1) there exists some locally convex space X, the space of the "analytic-like vectors", continuously and densely embedded in H, such that Pb{(1),(2)} is well-posed in X (in the sense that there exists a unique solution $u(t)$ in $W^{2,1}(0,T;X)$ for every v_0 and v_1 in X).

In its generality such a question has a negative answer. However, Th. 1 below gives an affirmative answer in a special case, which is sufficiently general to include Eq.(9) (at least if the coefficients are periodic in x) as well as some non-Kovalewskian equations such as

$$u_{tt} + c(t)\ \Delta_x^2 u = 0 \quad , \quad c(t) \geqslant 0 .$$

Before stating our result, we give some preliminary definitions.

Definition 1

A Banach scale $\{X_r\}$, $0 < r < \bar{r}$, is a family of Banach spaces (with norms $\| \ \|_r$) depending on a real parameter, such that

$$X_r \subseteq X_{r-\delta} \quad \underline{\text{and}} \quad \|v\|_{r-\delta} \leqslant \|v\|_r \quad , \ \underline{\text{if}} \quad 0 < \delta < r .$$

We also consider the locally convex inductive limit

$$X_{o^+} = \bigcup_{r>o} X_r$$

and the Fréchet spaces

$$X_{r^-} = \bigcap_{\delta>o} X_{r-\delta} \quad (r > 0), \quad X_\infty = \bigcap_{>o} X_r .$$

Definition 2

For any $v \in X_{o^+}$ the positive number

$$r_v = \text{Sup}\,\{ r > 0 \ : \ v \in X_r \}$$

is called the analyticity radius of v.

Definition 3

A Banach scale $\{X_r\}$ is said to be dense in itself if, for every $0 < \delta < r$, $X_{r+\delta}$ is dense in X_{r^-}.

Definition 4

A linear operator $A : X_\infty \longrightarrow X_\infty$ is said to have order m in the scale $\{X_r\}$ if there exists a positive constant λ s.t.

$$\|A v\|_{r-\delta} \leqslant \frac{\lambda}{\delta^m} \|v\|_r \qquad (0 < \delta < r) \tag{10}$$

for every $v \in X_\infty$.

Now we specialize these notions to the Banach scales generated by an operator.

Let B be a closed linear operator in the Hilbert space H and let us put

$$D(B^\infty) = \bigcap_{j=1}^{\infty} D(B^j) ,$$

(11) $\quad \|v\|_{r,B} = \sup_{j \geq 1} |B^j v|_H \frac{r^j}{j!}$.

Definition 5

The family of Banach spaces

$$X_r(B) = \{ v \in D(B^\infty) : \|v\|_{r,B} < \infty \} \qquad (r > 0),$$

equipped with the norm $\| \ \|_{r,B}$, is called the Banach scale generated by B .

Sometimes we shall refer to the elements of $X_{o^+}(B)$ as to the B-analytic vectors and to the analyticity radius in $\{X_r(B)\}$ (see Def. 2) as the B - analyticity radius.

Definition 6

A linear operator $A : D(B^\infty) \longrightarrow D(B^\infty)$ is said to have ω-order m with respect to B if there exist two postive constants K and Λ such that

(12) $\quad |B^j Av|_H \leqslant K(j+m)! \sum_{h=o}^{j+m} |B^h v|_H \frac{\Lambda^{j+m-h}}{h!}$

for every $j \in \mathbb{N}$ and $v \in D(B^\infty)$.

Remark 1

If an operator A has ω-order m with respect to B , then it has order m in the scale $\{X_r(B)\}_{r<\frac{1}{\Lambda}}$. More precisely, if (12) holds for some K and Λ , then (10) holds for $0 < \delta < r < \frac{1}{\Lambda}$ and $\lambda = \frac{K\, m!}{1-r\Lambda}$.

We can now state our result.

THEOREM 1 ([1])

Let (V,H,V') be a Hilbert triplet and $A(t) : V \to V'$ a family of linear operators satisfying (3), (4) and (5). Let moreover $B : V \to H$ be a linear operator such that

(13) the norms $|v|_V$ and $|v|_H + |Bv|_H$ are equivalent on V

and that the Banach scale $\{X_r(B)\}$ generated by B is dense in itself (see Def. 5 and Def. 3). Assume that, for every $j \in \mathbb{N}$ and $v \in D(B^\infty)$,

(14) $B^j A(\cdot)v$ is H-measurable on $[0,T]$,

(15)
$$\left| \begin{aligned} & |[A(t), B^j]v|_H \leqslant \\ & \leqslant \sqrt{\alpha(t)}(j+2)! \left[\langle A(t)B^j v, B^j v\rangle^{1/2} \frac{\Lambda}{(j+1)!} + \right. \\ & \left. + \sqrt{\alpha(t)} \sum_{h=0}^{j} |B^h v|_H \frac{\Lambda^{j+2-h}}{h!} \right] \end{aligned} \right.$$

(where $[\,,\,]$ denotes the commutator) and

(16) $$|A(t)v|_H \leqslant \beta(t) \, (|v|_H + |Bv|_H + |B^2 v|_H)$$

for some non-negative integrable functions $\alpha(t)$ and $\beta(t)$, and some constant $\Lambda \geqslant 0$.

Then Pb. $\{(1),(2)\}$ is well-posed in $X_{0^+}(B)$.

More precisely, if the initial date v_0 and v_1 belong to $X_{r_0}(B)$ for some $r_0 < \frac{1}{\Lambda}$, and we set

(17) $$r(t) = r_0 \exp\left[-\Lambda \left(1 + \frac{2}{\sqrt{1-r_0\Lambda}}\right) \int_0^t \sqrt{\alpha(s)}\, ds \right],$$

then there exists a unique solution u on $[0,T]$ such that the B-analyticity radii of $u(t)$ and $u'(t)$ are greater than or equal to $r(t)$.

Remark 2

The same conclusion of th. 1 remains valid if $A(t)$ is perturbed by any family $P(t)$ of operators with ω-order 1 with respect to B.

Remark 3

A consequence of (15) and (3) is the following inequality

$$(18) \qquad \left| [A(t),B^j]\, v \right|_H \leq \gamma(t)\, (j+2)! \sum_{h=o}^{j+1} |B^h v|_H \frac{\Lambda^{j+2-h}}{h!}$$

(for some function $\gamma(t)$), which is in particular satisfied whenever the commutator $[A(t),B]$ has ω-order 2 w.r. to B, uniformly in t.

In general, we cannot replace (15) by (18) in th. 1. We observe, however, that (18) and (15) are equivalent in the special case in which $A(t)$ satisfies the coerciveness condition (7).

Remark 4

When $[A(t),B] = 0$, condition (15) is trivially satisfied for $\alpha(t) \equiv 0$. In this case (17) reduces to $r(t) \equiv r_o$, i.e. the B-analyticity radius of $(u(t),u'(t))$ remains constant in time.

§ 2 - A PROOF OF TH. 1

Lemma 1 (local existence)

If $A(t)$ is any integrable family of linear operators of order m in a Banach scale $\{X_r\}$ (see Def. 4), the Cauchy problem

$$(19) \qquad \begin{cases} u^{(m)} + A(t)\, u = 0 & (0 \leq t \leq T) \\ u^{(j)}(0) = v_j & (j = 0,1,\ldots,m-1) \end{cases}$$

is locally well-posed in X_{o+}.

More precisely if, for every $v \in X_r$ and $0 < \delta < r$,

$$(20) \qquad A(\cdot)v \text{ is } X_{r-\delta}\text{-measurable on } [0,T]$$

and

$$(21) \qquad \|A(t)\, v\|_{r-\delta} \leq \frac{\lambda(t)}{\delta^m} \|v\|_r \quad \text{with} \quad \lambda \in L^1(0,T),$$

then, for $v_o v_1, \ldots, v_{m-1}$ in X_{r_o}, there is a unique solution $u(t)$ of Pb. (19) in any interval $[0,\tau[$ in which the function

$$(22) \qquad \tilde{r}(t) = r_o - \left(\frac{e}{(m-1)!}\right)^{\frac{1}{m}} \cdot \int_o^t \sqrt[m]{\lambda(s)}\, ds$$

remains positive. Moreover the analyticity radius of $u(t), u'(t), \ldots, u^{(m-1)}(t)$ is greater than or equal to $\tilde{r}(t)$.

(This Lemma is an extension of the Yamanaka-Ovciannikov theorem (see [11] and [10] to the case $m > 1$; for a proof including the estimate (22) of the analyticity radius, see [2]).

Lemma 2 (a priori estimate)

Let (V,H,V') and B be as in th. 1, and $A(t)$ a family of linear operators verifying (3),(4) (5) and (15).

Let $u \in W^{2,1}(0,T ; X_{r_o}(B))$ be any solution of Pb. {(1),(2)} with $r_o < \frac{1}{\Lambda}$, and let

$$e_j(t) = |B^j u(t)|_H^2 + |B^{j-1} u'(t)|_H^2 \qquad (j \geq 1) \tag{23}$$

be the energy of order j of u.

We have then, for any $\varepsilon > 0$,

$$\sum_{j=1}^{\infty} \sqrt{e_j(t)}\, \frac{(r(t)-\varepsilon)^{j+1}}{(j+1)!} \leq C_\varepsilon \sum_{j=1}^{\infty} \sqrt{e_j(0)}\, \frac{r_o^{j+1}}{(j+1)!} \tag{24}$$

where $r(t)$ is the function defined by (17) and C_ε a constant depending on $\varepsilon, A, \alpha, r_o, T$ and Λ.

Proof of Lemma 2(*)

In order to avoid some technicalities, we shall consider a rather special case, assuming that

$$A(t) \in C([0,T], \mathcal{L}(V,V')) \tag{25}$$

and that

(26) $\alpha(t)$ is C^1 and strictly positive on $[0,T]$

(where $\alpha(t)$ is the function appearing in (15).

(*) The proof which we give here is rather different from that given in [1] and it has the advantage that it also permits the treatment of the solvability in the Gevrey abstract spaces (see § 4).

For the general case, see [1].

Moreover we shall assume that

(27) $\qquad |v|_V = |v|_H + |Bv|_H \quad$ and $\quad \Lambda = 1$

(this is not actually restrictive). We have then in particular $r_o < 1$.

Let us now fix a family $\{A_j(t)\}$, $j = 1,2,\ldots$, of regular (in t), symmetric and coercive operators from V into V', and let us put

$$(28)\quad \left\{\begin{aligned} M_j(t) &\equiv M(A_j,t) = \|A_j'(t)\| \\ \nu_j(t) &\equiv \nu(A_j,t) = \operatorname*{Inf}_{v\neq o} \langle A_j(t)v,v\rangle . |v|_V^{-2} \\ \mu_j(t) &\equiv \mu(A_j,t) = \| A(t) - A_j(t) \| \\ \eta_j(t) &\equiv \eta(A_j,t) = \| (A(t) - A_j(t))^+ \| \end{aligned}\right.$$

where $\|S\|$ is the norm in $\mathcal{L}(V,V')$ and $\|S^+\|$ is equal to the Sup. of $\langle S v,v\rangle . |v|_V^{-2}$ for $v \neq 0$.

Finally let us define

$$(29)\quad E_j(t) = \langle A_j(t)\, B^{j-1}\, u(t)\, ,\, B^{j-1}\, u(t)\rangle + |B^{j-1}\, u'(t)|_H^2 ,$$

where $u(t)$ is the given solution of Pb. $\{(1), (2)\}$.

We have in particular

$$(30)\quad |B^{j-1}\, u|_V \leqslant \frac{\sqrt{E_j}}{\sqrt{\nu_j}}\ ,\ |B^{j-1}\, u'|_H \leqslant \sqrt{E_j}\ ,\ |B^{j-1}\, u'|_V \leqslant \sqrt{E_j} + \sqrt{E_{j+1}} .$$

If we apply B^{j-1} to any term of Eq. (1), we obtain the equation

$$B^{j-1}\, u'' + A_j\, B^{j-1}\, u = (A_j - A)\, B^{j-1}\, u + [A,B^{j-1}]\, u .$$

Therefore, by differentiating in (29), we get

$$E_j' = \langle A_j'\, B^{j-1}\, u,\, B^{j-1}\, u\rangle + 2\,\mathrm{Re} \langle (A_j - A)\, B^{j-1}\, u\, ,\, B^{j-1}\, u'\rangle$$
$$+ 2\,\mathrm{Re} \langle [A,B^{j-1}]\, u\, ,\, B^{j-1}\, u'\rangle ,$$

and hence, using (28) and (30),

(31) $$E'_j \leq \frac{M_j}{\nu_j} E_j + 2\frac{\mu_j}{\sqrt{\nu_j}} \sqrt{E_j}\,(\sqrt{E_j} + \sqrt{E_{j+1}}) + 2\,|[A,B^{j-1}]u|_H \sqrt{E_j}\,.$$

To estimate the last term of (31) we use assumption (15), i.e. the inequality

(32) $$|[A,B^{j-1}]u|_H \leq \sqrt{\alpha}(j+1)\ \langle A\,B^{j-1}u, B^{j-1}u\rangle^{1/2} + \alpha(j+1)!\sum_{h=0}^{j-1}\frac{|B^h u|_H}{h!}\,.$$

Since

$$\langle AB^{j-1}u, B^{j-1}u\rangle^{1/2} = \left(\langle A_j\,B^{j-1}u, B^{j-1}u\rangle + \langle (A-A_j)B^{j-1}u, B^{j-1}u\rangle\right)^{1/2}$$

$$\leq \left((1+\frac{\eta_j}{\nu_j})\,E_j\right)^{1/2}$$

$$\leq\ 1+\frac{\sqrt{\eta_j}}{\sqrt{\nu_j}}\ \sqrt{E_j}$$

and $|B^h u|_H \leq \sqrt{E_h}\ (\sqrt{\nu_h})^{-1}$ for $h \geq 2$, while $|u|_H + |Bu|_H \equiv$

$\equiv\ |u|_V \leq \sqrt{E_1}\ \ (\sqrt{\nu_1})^{-1}$, (32) gives :

(33) $$|[A,B^{j-1}]u|_H \leq \sqrt{\alpha}(j+1)\ (1+\frac{\sqrt{\eta_j}}{\sqrt{\nu_j}})\sqrt{E_j} + \alpha(j+1)!\sum_{h=1}^{j-1}\frac{\sqrt{E_h}}{\sqrt{\nu_h}\,h!}\,.$$

Introducing (33) in (31) and dividing by $\sqrt{E_j}$ we obtain (using the identity $\sqrt{E_j}' = E_j'(2\sqrt{E_j})^{-1}$)

(34) $$\sqrt{E_j}' \leq \left(\frac{M_j}{2\nu_j} + \frac{\mu_j}{\sqrt{\nu_j}}\right)\sqrt{E_j} + \frac{\mu_j}{\sqrt{\nu_j}}\ \sqrt{E_{j+1}}\ +$$
$$+\sqrt{\alpha}(j+1)(1+\frac{\sqrt{\eta_j}}{\sqrt{\nu_j}})\ \sqrt{E_j} + \alpha(j+1)!\sum_{h=1}^{j-1}\frac{\sqrt{E_h}}{\sqrt{\nu_h}\,h!}\,.$$

The presence of $\sqrt{E_{j+1}}$ in the right hand side of (34) prevents us from deriving an estimate of $\sqrt{E_j}$ in terms of $\sqrt{E_1}(0),\ldots,\sqrt{E_j}(0)$. Thus we introduce the <u>infinite order energy</u> function

(35) $$\mathcal{E}(t) = \sum_{j=1}^{\infty}\sqrt{E_j(t)}\ \frac{\rho(t)^{j+1}}{(j+1)!}\quad,$$

where $\rho(t)$ is a regular positive function, which will be chosen later on, such that $\rho(t) \leqslant r_o$.

By differentiating term by term in (35) and using (34), we have

$$\mathcal{E}' = \sum_{j=1}^{\infty} \left(\sqrt{E_j}\, \frac{\rho^j}{j!}\, \rho' + \sqrt{E_j'}\, \frac{\rho^{j+1}}{(j+1)!} \right)$$

$$\leqslant \sum_{j=1}^{\infty} \sqrt{E_j}\, \frac{\rho^j}{j!}\, \rho' + \sum_{j=1}^{\infty} \sqrt{E_j} \left(\frac{M_j}{2\nu_j} + \frac{\mu_j}{\sqrt{\nu_j}} \right) \frac{\rho^{j+1}}{(j+1)!} +$$

$$\sum_{j=1}^{\infty} \sqrt{E_{j+1}}\, \frac{\mu_j}{\sqrt{\nu_j}}\, \frac{\rho^{j+1}}{(j+1)!} + \sqrt{\alpha} \sum_{j=1}^{\infty} \sqrt{E_j} \left(1 + \frac{\sqrt{\eta_j}}{\sqrt{\nu_j}} \right) \frac{\rho^{j+1}}{j!} +$$

$$+\ \alpha \,.\, \sum_{j=1}^{\infty} \rho^{j+1} \sum_{h=1}^{j-1} \frac{\sqrt{E_h}}{\sqrt{\nu_h}\, h!} \,.$$

By changing the order of summation, we can calculate the last term of this inequality as

$$\alpha \,.\, \sum_{j=1}^{\infty} \rho^{j+1} \sum_{h=1}^{j-1} \frac{\sqrt{E_h}}{\sqrt{\nu_h}\, h!} = \alpha\, \frac{\rho^2}{1-\rho} \sum_{j=1}^{\infty} \frac{\sqrt{E_j}}{\sqrt{\nu_j}}\, \frac{\rho^j}{j!}$$

so that we obtain the inequality

$$\mathcal{E}' \leqslant \sum_{j=1}^{\infty} \sqrt{E_j}\, \frac{\rho^j}{j!}\, \psi_j \tag{36}$$

where

$$\psi_j = \rho' + \left(\frac{M_j}{2\nu_j} + \frac{\mu_j}{\sqrt{\nu_j}} \right) \frac{\rho}{j+1} + \frac{\mu_{j-1}}{\sqrt{\nu_{j-1}}} + \sqrt{\alpha} \left(1 + \frac{\sqrt{\eta_j}}{\sqrt{\nu_j}} \right) \rho + \frac{\alpha}{\sqrt{\nu_j}}\, \frac{\rho^2}{1-\rho} \,. \tag{37}$$

Now, in view of (24) we choose the operators $A_j(t)$ and the function $\rho(t)$ (dependent on ε) in such a way that $\psi_j \leqslant 0$ for j sufficiently large with respect to ε.

More precisely, we define $\rho(t) \equiv \rho_\varepsilon(t)$ equal to the solution of the problem

$$\rho_\varepsilon' + \sqrt{\alpha}\left(1 + \frac{2}{\sqrt{1-r_o}}\right) \rho_\varepsilon = -\varepsilon \qquad \text{on } [0,T] \,, \tag{38}$$

$$(39)\qquad \rho_\varepsilon(0) = r_o .$$

If we compare Eq.(38) with the differential equation $r' + \sqrt{\alpha}(1+2(1-r_o)^{-1/2})\, r = 0$ satisfied by $r(t)$ (see (17) with $\Lambda = 1$) we see, taking (26) into account, that

$$(40)\qquad 0 < \frac{1}{2}\, r(T) \leq r(t) - C.\varepsilon \leq \rho_\varepsilon(t) \leq r(t) \leq r_o < 1 ,$$

for some constant $C = C(\alpha, r_o, T)$ and ε sufficiently small.

Now let us define the function

$$(41)\qquad \nu_\varepsilon(t) = \alpha(t)\, \frac{\rho_\varepsilon^2(t)}{1-\rho_\varepsilon(t)} ,$$

and the operators

$$(42)\qquad A_{\varepsilon,j}(t) = (A^*\varphi_{\delta_j})(t) + \nu_\varepsilon(t)(B^*B + Id) ,$$

where $B^*: H \to V'$ is the dual operator of $B : V \to H$, $\varphi_{\delta_j}(t)$ is a family of <u>Friedrichs mollifiers</u> with support in $[-\delta_j, \delta_j]$, $A*\varphi_{\delta_j}$ is the convolution-product of φ_{δ_j} with A (after the extension : $A(t) \equiv A(0)$ for $t < 0$ and $A(t) \equiv A(T)$ for $t > T$) and $\{\delta_j\}$ is a sequence decreasing to zero of positive numbers which will be chosen in an appropriate way (see (51) below).

The parameters introduced by (28) can be estimated as follows :

$$(43)\qquad \begin{cases} M_j \equiv M(A_{\varepsilon,j}) \leq \dfrac{\omega(\delta_j)}{\delta_j} + |\nu_\varepsilon'(t)| \\[2ex] \eta_j \equiv \eta(A_{\varepsilon,j}) \leq \omega(\delta_j) \\[2ex] \mu_j \equiv \mu(A_{\varepsilon,j}) \leq \omega(\delta_j) + \nu_\varepsilon(t) \\[2ex] \nu_j \equiv \nu(A_{\varepsilon,j}) \geq \nu_\varepsilon(t) , \end{cases}$$

where $\omega(\delta)$ denotes the <u>modulus of continuity</u> of $A(t)$, i.e.

(44) $$\omega(\delta) = \sup_{\substack{|\tau|\leqslant\delta \\ 0\leqslant t\leqslant T}} \| A(t+\tau) - A(t) \| \quad .$$

We observe that $\omega(\delta)$ is infinitesimal for $\delta \to 0$ in virtue of (25). Moreover, from (26), (41) and (40) we have

(45) $$0 \leqslant \nu_o \leqslant \nu_\varepsilon(t) \leqslant \frac{1}{\nu_o}, \; |\nu_\varepsilon'(t)| \leqslant \frac{1}{\nu_o} \; , \qquad \text{on } [0,T] \; ,$$

for some positive constant $\nu_o \equiv \nu_o(\alpha, r_o, T)$ and ε sufficiently small.

Let us denote by $E_{\varepsilon,j}$ the j-energy corresponding to the operator $A_{\varepsilon,j}$, i.e

(46) $$E_{\varepsilon,j} = \langle A_{\varepsilon,j} B^{j-1} u, B^{j-1} u \rangle + |B^{j-1} u'|_H^2 \; ,$$

and let us consider the corresponding infinite order energy

(47) $$\mathcal{E}_\varepsilon = \sum_{j=1}^{\infty} \sqrt{E_{\varepsilon,j}} \, \frac{\rho_\varepsilon^{j+1}}{(j+1)!} \; .$$

Therefore (36) and (37), together with (43), give

(48) $$\mathcal{E}'_\varepsilon \leqslant \sum_{j=1}^{\infty} \sqrt{E_{\varepsilon,j}} \, \frac{\rho_\varepsilon^j}{j!} \, \psi_{\varepsilon,j}$$

with

$$\psi_{\varepsilon,j} \leqslant \rho'_\varepsilon + \left(\frac{\omega(\delta_j)}{2\nu_\varepsilon \delta_j} + \frac{|\nu'_\varepsilon|}{2\nu_\varepsilon} + \frac{\omega(\delta_j)}{\sqrt{\nu_\varepsilon}} + \sqrt{\nu_\varepsilon} \right) \frac{\rho_\varepsilon}{j+1} +$$

$$+ \left(\frac{\omega(\delta_{j-1})}{\sqrt{\nu_\varepsilon}} + \sqrt{\nu_\varepsilon} \right) + \sqrt{\alpha} \left(1 + \frac{\sqrt{\omega(\delta_j)}}{\sqrt{\nu_\varepsilon}} \right) \rho_\varepsilon + \frac{\alpha}{\sqrt{\nu_\varepsilon}} \cdot \frac{\rho_\varepsilon^2}{1-\rho_\varepsilon} \; .$$

We write the last inequality in the form

(49) $$\psi_{\varepsilon,j} \leqslant \rho'_\varepsilon + \left(\sqrt{\alpha}\, \rho_\varepsilon + \sqrt{\nu_\varepsilon} + \frac{\alpha}{\sqrt{\nu_\varepsilon}} \cdot \frac{\rho_\varepsilon^2}{1-\rho_\varepsilon} \right) + \Phi_{\varepsilon,j}$$

where we put, for the sake of brevity ,

(50) $$\Phi_{\varepsilon,j} = \left(\frac{\omega(\delta_j)}{2\nu_\varepsilon \delta_j} + \frac{|\nu_\varepsilon'|}{2\nu_\varepsilon} + \frac{\omega(\delta_j)}{\sqrt{\nu_\varepsilon}} + \sqrt{\nu_\varepsilon} \right) \frac{\rho_\varepsilon}{j+1} + \frac{\omega(\delta_{j-1})}{\sqrt{\nu_\varepsilon}} + \sqrt{\alpha}\, \frac{\sqrt{\omega(\delta_j)}}{\sqrt{\nu_\varepsilon}}\, \rho_\varepsilon$$

Now we have, from (41), (40) and (38),

$$\rho'_\varepsilon + \sqrt{\alpha}\,\rho_\varepsilon + \sqrt{\nu_\varepsilon} + \frac{\alpha}{\sqrt{\nu_\varepsilon}}\,\frac{\rho_\varepsilon^2}{1-\rho_\varepsilon} =$$

$$= \rho'_\varepsilon + \sqrt{\alpha}\,\rho_\varepsilon + 2\sqrt{\alpha}\,\frac{\rho_\varepsilon}{\sqrt{1-\rho_\varepsilon}}$$

$$\leqslant \rho'_\varepsilon + \sqrt{\alpha}\,\rho_\varepsilon + 2\sqrt{\alpha}\,\frac{\rho_\varepsilon}{\sqrt{1-r_o}}$$

$$= -\varepsilon .$$

On the other hand, if we choose the sequence $\{\delta_j\}$ in such a way that

$$\text{(51)}\qquad \frac{\omega(\delta_j)}{\delta_j}\cdot\frac{1}{j+1} \longrightarrow 0 \quad , \quad \text{for } j \longrightarrow \infty,$$

(which is always possible), we get from (50), using (45), (40) and (26), that

$$\Phi_{\varepsilon,j} \longrightarrow 0 \ \text{ in } \ C([0,T]), \text{ for } j \to \infty .$$

In conclusion, going back to (49), we have

$$\psi_{\varepsilon,j} \leqslant -\varepsilon + \Phi_{\varepsilon,j} \leqslant \tilde{c}_\varepsilon \qquad \text{for every } j ,$$

and

$$\psi_{\varepsilon,j} \leqslant 0 \qquad \text{for } j \geqslant j_\varepsilon \ ;$$

so that (48) gives

$$\mathcal{E}'_\varepsilon \leqslant \sum_{j=1}^{j_\varepsilon} \sqrt{E_{\varepsilon,j}}\,\frac{\rho_\varepsilon^j}{j!}\,\psi_{\varepsilon,j} \leqslant \tilde{C}_\varepsilon \sum_{j=1}^{j_\varepsilon} \sqrt{E_{\varepsilon,j}}\,\frac{\rho_\varepsilon^j}{j!}$$

$$\leqslant \tilde{C}_\varepsilon\,\frac{(j_\varepsilon+1)}{\rho_\varepsilon} \sum_{j=1}^{j_\varepsilon} \sqrt{E_{\varepsilon,j}}\,\frac{\rho_\varepsilon^{j+1}}{(j+1)!} \leqslant \approx{C}_\varepsilon \cdot \mathcal{E}_\varepsilon ,$$

i.e.

$$\text{(52)}\qquad \mathcal{E}_\varepsilon(t) \leqslant \overset{\approx}{\tilde{C}}_\varepsilon \cdot \mathcal{E}_\varepsilon(0) \qquad \text{on} \quad [0,T] ,$$

where $\tilde{C}_\varepsilon$, $\overset{\approx}{C}_\varepsilon$ and $\overset{\approx}{\tilde{C}}_\varepsilon$ are positive constants depending on $\varepsilon, A, \alpha, r_o$ and T.

From (52) we deduce (24), taking (40) and (45) into account.

Conclusion of the proof of Th. 1

Using (15) and (16) we can see that $A(t)$ has ω-order 2 with respect to B, more precisely that $A(t)$ satisfies (12) with $K \equiv C \cdot (\alpha(t) + \beta(t))$ (for some constant C). Hence (21) holds for $m = 2$ (see Rem. 1). On the other hand we can prove that (14), together with (15) and (16), gives (20).

Therefore we are in the position to apply Lemma 1, which ensures a local solution for Pb. $\{(1),(2)\}$.

Now, using the "a priori" estimate of Lemma 2 and the fact that $\{X_r(B)\}$ is dense in itself (see Def. 3), we see that any local solution of our Problem can be continued to the whole time-interval $[0,T]$.

This concludes the proof of Th. 1 .

§ 3 - APPLICATIONS

Corollary 1 - (Second order weakly hyperbolic P.D.E.)

Let us consider the Cauchy problem for Eq.(9) where $c(x,t)$ is analytic in x and measurable in t and

(53) $$|D_x^j\ c(x,t)| \leqslant K(t)\ \Lambda^j\ j! \qquad (j \in \mathbb{N})$$

for some positive constant Λ and some K in $L^1(0,T)$.

Then

a) if $c(x,t)$ is 2π-periodic in x, the Problem is well-posed in the space $\mathscr{A}_{2\pi}(\mathbb{R}_x)$ of the analytic 2π-periodic functions.

b) If $c(x,t)$ belongs to $L^1(0,T;L^\infty(\mathbb{R}_x))$, the Problem is well-posed in the space $\mathscr{A}_{L^2}(\mathbb{R}_x)$ consisting of the analytic functions $v(x)$ such that

$$\|D^j v\|_{L^2(\mathbb{R}_x)} \leqslant C\ \Lambda^j\ j! \qquad (j \in \mathbb{N})$$

for some C and Λ (depending on v).

Proof

Let us apply Th. 1 with

$$B = D_x \ , \quad A(t) = -D_x(c(x,t)\ D_x)$$

and, in the Case (a).

$$H = L^2(0,2\pi) \quad , \quad V = \{v \in H^1_{loc} \ : \ v \ \text{ is } \ 2\pi\text{-periodic}\} \ ;$$

while in the Case (b)

$$H = L^2(\mathbb{R}_x) \quad , \ V = H^1(\mathbb{R}_x) \ .$$

We have then

$$[A(t), B^j] = j . D_x c(x,t) . D_x^{j+1} + \text{lower order terms} \ ;$$

hence, using the inequality

$$(54) \qquad |f'(x)|^2 \leqslant 2 \, \|f''\|_{L^\infty} . \, f(x) \quad , \quad \text{if} \quad f \geqslant 0 \, ,$$

(which is due to G. GLAESER ([6])) and also (53) for $j = 2$, we obtain (15) with $\alpha(t) = 4\,K(t)$.

Remark 5

This result was proved, in the general case of arbitrary analytic initial data and coefficients, by E. JANNELLI (see [7]).

Remark 6

Corollary 1 can be extended to the case of several space-variables by applying instead of Th. 1 a slight generalisation of this theorem to the case of Banach scales generated by a n-tuple of operators $B_1, \ldots, B_n$ commuting among then (see [1]). The basic assumption is now a condition similar to (15), which in the concrete cases may be obtained using an extension of inequality (54) which is due to O.A. OLEINIK (Lemma 4 of [9]).

Corollary 2 (boundary conditions)

Let us consider the Cauchy problem for the equation

$$(55) \qquad u_{tt} - c(t) \ \Delta_x u = 0 \quad , \quad \text{on} \ \ \Omega \times [0,T] \ ,$$

where $c(t)$ is a non-negative, integrable function on $[0,T]$ and Ω an open subset of $\mathbb{R}^n_x$.

Then the Problem is well-posed in the space $\mathscr{A}(\sqrt{-\Delta_x}\,;\Omega)$ which consists of the analytic functions $v(x)$ on Ω such that

(56) $\Delta_x^j v \in W_o^{1,2}(\Omega)$

(57) $\| \Delta_x^j v \|_{L^2(\Omega)} \leqslant C \, \Lambda^{2j} \, j!^2$

for every $j \in \mathbf{N}$ (We recall that, if Ω is bounded and regular, a function $v(x)$ which satisfies (56) for every j verifies also (57) if and only if v is analytic on some neighborhood of Ω).

Proof

Apply Th. 1 with $B = \sqrt{-\Delta_x}$ and $A(t) = -c(t)\Delta_x$. Condition (15) is trivial since $[A(t), B] = 0$.

Corollary 3 (non Kovalewskian equations)

The Cauchy problem for the equation

$$u_{tt} + c(t)\Delta_x^2 u = 0 \qquad (x \in \mathbf{R}^n, \ t \in [0,T])$$

where $c(t)$ is $\geqslant 0$ and integrable on $([0,T])$, is well-posed in the space $\mathscr{G}^{1/2}(\mathbf{R}_x^n)$ consisting of the functions $v(x)$ analytic and such that

$$\| D_x^\alpha v \|_{L^2(\mathbf{R}_x^n)} \leqslant C \Lambda^{|\alpha|} \sqrt{\alpha}\,! \quad , \quad \forall \alpha$$

Proof

Apply Th. 1 with $B = \Delta_x$ and $A(t) = c(t)\Delta_x^2$.

§ 4 - THE GEVREY WELL-POSEDNESS

In [3] and [8] it was proved that the Cauchy problem for Eq.(8), or Eq.(9), is well-posed also in the Gevrey space $\mathscr{G}^s(\mathbf{R}_x)$, $s > 1$, provided that the coefficients of the equation belong to $C^{o,\vartheta}(0,T)$, or to $C^{o,\vartheta}(0,T\,;\mathscr{G}^s(\mathbf{R}_x))$, for ϑ sufficiently large depending on s .

In order to extend such a result to the abstract case, we must first define the Banach scale $\{X_r^s(B)\}$, $r > 0$, of the B-Gevrey vectors with order s , where s is a fixed real number > 1 . The Banach space $X_r^s(B)$ is defined in a similar way to $X_r(B)$ (see Def. 5), starting from the norm

$$\|v\|_{r,s,B} = \sup_{j \geqslant o} |B^j v|_H \frac{r^j}{j!^s} .$$

Moreover, to treat the differential operators with order $\underline{m}$ having Gevrey coefficients, we extend Def. 6 by saying that a linear operator $A : D(B^\infty) \to D(B^\infty)$ has $\underline{\omega_s\text{-order } m \text{ with respect to } B}$ whenever

$$|B^j A v|_H \leqslant K(j+m)!^s \sum_{h=o}^{j+m} |B^h v| \frac{\Lambda^{j+m-h}}{h!^s}$$

for every $j \in \mathbb{N}$ and $v \in D(B^\infty)$.

It is important to notice that any operator, with ω_s-order $\underline{m}$ w.r. to B, has order $\underline{m\,s}$ in the scale $X_r^s(B)$ (see Def. 4) ; so that the equation

$$u^{(k)} + A u = 0$$

is Kovalewskian in this scale only if $\underline{k \geqslant m\,s}$ (see [C]).

In particular, Eq. (1) is non-Kovalewskian in the scale $\{X_r^s(B)\}$ for $s > 1$, and Lemma 1 is no longer true.

On the other hand it is possible to extend Lemma 2 to the case $s > 1$, by proving an "a priori" estimate such as (24) for the X_r^s-norms of any solution to Eq. (1), provided that A(t) satisfies conditions (4), (5), (15) and, in addition, an appropriate Hölder condition. Such an "a priori" estimate holds true even if assumption (15) is weakened by

$$|[A,B^j]v|_H \leqslant$$

$$\leqslant \sqrt{\alpha(t)}\,(j+2) < A\,B^j v, B^j v >^{1/2} \Lambda + \alpha(t)\,(j+2)(j+1)!^s \sum_{h=o}^{j} |B^h v| \frac{\Lambda^{j+2-h}}{h!} .$$

For instance, we have the "a priori" estimate of $\|u(t)\|_{r,s,B}$ if A(t) is V-coercive (i.e. (7) holds) and

(58) $\quad A \in C^{o,\vartheta}(O,T ; \mathcal{L}(V,V'))$ for $\vartheta > 1 - \frac{1}{s}$.

To prove this fact we need only to modify the last part of the proof of Lemma 2 by taking

$$A_j = A * \varphi_{\delta_j} \quad \text{and} \quad \delta_j = (\tfrac{1}{j})^s \qquad (\nu_\varepsilon \equiv 0)$$

(cf. (42) and (51)) and remarking that now

$$\omega(\delta_j) \leqslant C \,.\, \delta_j^{\vartheta}$$

by virtue of (58).

Finally, we observe that if we want to use the "a priori" estimate to obtain the global well-posedness of Pb {(1), (2)} in $X^s_{o^+}(B)$, we must now assume some additional conditions which make up for the lack of the local existence. For instance, a compactness condition on the operator B^{-1} could permit us to apply the Riesz-Galerkin method and hence to construct a sequence of approximating solutions for Pb {(1), (2)} .

REFERENCES

[1] A. AROSIO - S. SPAGNOLO : Global existence for abstract evolution equations of weakly hyperbolic type, Pubblicazioni del Dip. de Mat., Universita di Pisa.

[2] L. CARDOSI : Evolution equations in scales of abstract Gevrey spaces, in preparation.

[3] F. COLOMBINI - E. DE GIORGI - S. SPAGNOLO : Sur les équations hyperboliques avec des coefficients qui ne dépendent que du temps, Ann. Scu. Norm. Sup. Pisa, 6 (1979), 511-559.

[4] F. COLOMBINI - S. SPAGNOLO : Second order hyperbolic equations with coefficients real analytic in the space variables and discontinuous in time, J. Analyse Math. 38 (1980), 1-33.

[5] F. COLOMBINI - S. SPAGNOLO : An example of a weakly hyperbolic Cauchy problem not well posed in C^∞, Acta Math. 148 (1982), 243-253.

[6] G. GLAESER : Racine carrée d'une fonction différentiable, Ann. Inst. Fourier 13 (1963), 203-210.

[7] E. JANNELLI : Weakly hyperbolic equations of second order with coefficients real analytic in space variables; Comm. in P.D.E. (1962), 537-558.

[8] T. NISHITANI : Sur les équations hyperboliques à coefficients qui sont höldériens en t et de classe de Gevrey en x, Bull. Sci. Math. 107 (1983), 113-138.

[9] O.A. OLEINIK : On linear equations of second order with non-negative characteristic form, Mat. Sbornik, 69 (1966), 111-140. English transl. : Ann. Math. Soc. (2) 65 , 167-199.

[10] L.V. OVSJANNIKOV : A singular operator in a scale of Banach spaces, Dokl. Akad. Nauk SSSR 163 (1965), 819-822. English transl. Soviet Math. Doklady 6 (1965).

[11] T. YAMANAKA : Note on Kowaleskaja's system of partial differential equations, Comment. Math. Univ. Saint Paul 9 (1960), 7-10.

S. SPAGNOLO
Scuola Normale Superiore
Piazza dei Cavalieri
PISA (Italy)

L TARTAR

Estimations fines des coefficients homogénéisés

Historique

L'homogénéisation est maintenant une théorie mathématique bien développée avec des méthodes considérées comme classiques ; pourtant il y a moins de dix ans rien ne semblait présager cet essor. Autour de E. DE GIORGI l'école italienne s'intéressait depuis déja quelques années à des problèmes de convergence d'opérateurs [3, 8, 11, 12] ; en France j'avais avec F. MURAT, considéré des questions analogues en relation avec des problèmes d'optimisation de domaines et déja le problème des estimations optimales dont je vais vous parler était posé, mais nous ne pouvions y donner qu'une réponse partielle [13] . Ce n'est qu'en 1975 que j'apprenais le terme d'homogénéisation et, aiguilloné par l'intérêt que J.L. LIONS portait maintenant à la question, je travaillais à mettre au point les méthodes générales que j'exposais dans mon cours Peccot en 1977.

La raison du développement rapide de la théorie a certainement été la multiplicité des problèmes de type homogénéisation dans les sciences de l'ingénieur mais cette multiplicité a malheureusement canalisé beaucoup d'efforts dans une phase d'exploitation où peu d'idées nouvelles apparaissaient. Mon problème d'estimations optimales des coefficients homogénéisés n'intéressait toujours personne, l'existence de formules dites explicites apparaissant dans les problèmes à structure périodique (qui monopolisaient l'attention des chercheurs français) laissait croire à beaucoup que le problème était facile, sinon résolu.

Mon message que la clé des phénomènes d'homogénéisation était un lemme de compacité par compensation associé à la construction de fonctions test n'était pas très bien passé et c'est pourtant encore en utilisant la compacité par compensation que j'imaginais une méthode pour donner des conditions nécessaires satisfaites par les coefficients homogénéisés [15] ; l'application de cette méthode n'était pas aisée et c'est seulement lors d'un séjour à New-York en 1980 que je trouvais comment l'utiliser et avec une construction de sphères emboitées de Hashin et Shtrikman [4] que G. PAPANICOLAOU me signalait alors, cela permettait de clarifier le cas des mélanges isotropes. Immédiatement, avec F. MURAT, nous tentâmes de traiter le cas non isotrope en essayant d'utiliser des ellipsoïdes à la place des sphères dans la construction d'un matériau test. E.L. FRAENKEL nous apporta une aide substantielle pour comprendre le cas de la dimension 2 et nous suggera l'usage des coordonnées ellipsoïdales pour le cas général ; nous n'eûmes pas le courage de suivre cette idée mais après avoir résolu le problème nous vîmes que nous avions redécouvert des formules analogues.

Le résultat exposé à New-York en juin 1981, a été ensuite simplifié par P. BRAIDY et D. POUILLOUX [2] qui montrèrent que la construction des matériaux feuilletés permettait de fabriquer les cas les plus généraux. Des résultats analogues en dimension 2 ont aussi été obtenus indépendamment par K.A. LURIE et A.V. CHERKAEV [6] .

Si l'utilisation des bornes pour des problèmes d'optimisation de domaines a été une motivation commune à K.A. LURIE - A.V. CHERKAEV [7] , R. KOHN-G. STRANG [5] , F. MURAT - L. TARTAR [10] , il faut noter que le problème des bornes a intéressé des physiciens depuis plus longtemps, et comme je ne peux ici parler de l'ensemble de leurs résultats je me contenterai de signaler le cours de D. BERGMAN [1] .

I. Le problème

Dans un ouvert Ω de $\mathbb{R}^N$ on considère une équation du type

$$(1) \qquad -\sum_{i,j} \frac{\partial}{\partial X_i}\left(A^{\varepsilon}_{ij}(x)\frac{\partial u^{\varepsilon}}{\partial X_j}\right) = f$$

où les coefficients A^{ε}_{ij} vérifient

$$(2) \qquad \begin{cases} A^{\varepsilon}_{ij} = A^{\varepsilon}_{ji} \in L^{\infty}(\Omega) \quad \forall i,j \\ \alpha|\xi|^2 \leqslant \sum_{i,j} A^{\varepsilon}_{ij}(x)\,\xi_i\,\xi_j \leqslant \beta|\xi|^2 \quad \text{p.p. } x \in \Omega\,,\ \forall\,\xi \in \mathbb{R}^N. \\ 0 < \alpha \leqslant \beta < +\infty \end{cases}$$

La matrice $A^{\varepsilon}(x) = \left(A^{\varepsilon}_{ij}\right)_{ij}$ dépend d'un indice ε et on s'intéresse à des problèmes de convergence en ε (dans la pratique ε est une longueur très petite devant le diamètre de Ω et le problème précédent est un modèle mathématique pour décrire les propriétés macroscopiques d'un matériau dont on connait la structure microscopique).

Pour pouvoir passer à la limite dans (1) on est amené à introduire sur les matrices $A^{\varepsilon}(\cdot)$ une topologie adaptée. On peut comme dans les travaux de DE GIORGI, MARINO, SPAGNOLO associer à (1) des conditions aux limites et définir la G-convergence comme convergence faible des opérateurs $f \to u^{\varepsilon}$ et ce procédé, équivalent à le minimisation de fonctionnelles quadratiques, conduit naturellement à la notion de Γ-convergence introduite par DE GIORGI pour le cas de fonctionnelles non quadratiques ; on peut aussi et c'est l'approche que j'ai suivie avec F. MURAT, travailler dans un cadre de systèmes d'équations aux dérivées partielles et introduire la notion analogue de H-convergence suivante (si dans le cas symétrique la définition est assez voisine, l'avantage de cette méthode est de mettre en évidence des quantités physiques importantes).

Définition

On dira qu'une suite de matrices A^{ε} H-converge vers A^{o} (on notera $A^{\varepsilon} \xrightarrow{H} A^{o}$) si pour toute suite E^{ε}, D^{ε} vérifiant

$$(3) \qquad \begin{cases} E^{\varepsilon} \longrightarrow E^{o} \quad \text{dans } L^2(\Omega)^N \quad \text{faible} \\ \dfrac{\partial E^{\varepsilon}_i}{\partial X_j} - \dfrac{\partial E^{\varepsilon}_j}{\partial X_i} \in \text{compact de } H^{-1}_{loc}(\Omega) \qquad \forall i,j \end{cases}$$

$$(4)\quad \begin{cases} D^{\varepsilon} \longrightarrow D^{o} \quad \text{dans} \quad L^2(\Omega)^N \quad \text{faible} \\ \sum_j \dfrac{\partial D^{\varepsilon}_j}{\partial X_j} \in \text{compact de } H^{-1}_{loc}(\Omega) \end{cases}$$

$$(5)\quad D^{\varepsilon} = A^{\varepsilon} E^{\varepsilon}$$

on puisse déduire (6) $D^o = A^o E^o$. ■

La propriété clé est d'utiluser le fait que (3) et (4) impliquent $E^{\varepsilon} \cdot D^{\varepsilon} \longrightarrow E^o \cdot D^o$ au sens des distributions ; cette propriété, premier exemple de la compacité par compensation que nous allons retrouver plus loin, alliée à la construction de fonctions test comme solutions d'un problème aux limites associé permet de montrer que l'ensemble (2) est compact pour la H-convergence.

Une question très importante est ensuite de savoir relier les oscillations du champ E^{ε} à sa valeur macroscopique E^o et d'introduire une matrice $\mathbb{P}^{\varepsilon}$ qui a les propriétés suivantes

$$(7)\quad \begin{cases} \mathbb{P}^{\varepsilon} \longrightarrow I \quad \text{dans} \quad L^2(\Omega)^{N^2} \quad \text{faible} \\ \dfrac{\partial \mathbb{P}^{\varepsilon}_{ij}}{\partial X_k} - \dfrac{\partial \mathbb{P}^{\varepsilon}_{ik}}{\partial X_j} \in \text{compact de } H^{-1}_{loc}(\Omega) \quad \forall\, i,j,k \end{cases}$$

$$(8)\quad \begin{cases} Q^{\varepsilon} = A^{\varepsilon}\,\mathbb{P}^{\varepsilon} \longrightarrow A^o \quad \text{dans} \quad L^2(\Omega)^{N^2} \quad \text{faible} \\ \sum_j \dfrac{\partial Q^{\varepsilon}_{ij}}{\partial X_j} \in \text{compact de } H^{-1}_{loc}(\Omega) \quad \forall i . \end{cases}$$

On montre alors que (3) (4) (5) impliquent

$$(9)\quad \begin{cases} E^{\varepsilon} - \mathbb{P}^{\varepsilon} E^o \longrightarrow 0 \quad \text{dans} \quad L^1_{loc}(\Omega)^N \quad \text{fort} \\ D^{\varepsilon} - Q^{\varepsilon} E^o \longrightarrow 0 \quad \text{dans} \quad L^1_{loc}(\Omega)^N \quad \text{fort} . \end{cases}$$

Le problème qui nous intéresse ici est de relier A^o à des informations simples sur A^{ε} : par exemple; connaissant les limites faibles de fonctions de A^{ε}, que peut-on en déduire sur A^o ?

La construction de la matrice $\mathbb{P}^{\varepsilon}$ se fait, au cours de la démonstration de l'existence de A^o pour une sous-suite, à l'aide de solutions d'un problème aux limites adapté [9], [14].

II. Caractérisation des mélanges de deux matériaux isotropes

On suppose que

$$(10) \quad \begin{cases} A^{\varepsilon}(x) = C^{\varepsilon}(x)\, I \ ; \quad C^{\varepsilon}(x) = \{ {}^{\alpha}_{\beta} \quad \text{p.p.} \\ C^{\varepsilon} \longrightarrow \theta\,\alpha + (1-\theta)\beta \quad \text{dans } L^{\infty}(\Omega) \text{ faible} \star \end{cases}$$

$\theta(x)$ est la proportion locale du matériau de caractéristique α. Nous aurons besoin des notations suivantes

$$(11) \quad \begin{cases} \mu_{+}(\theta) = \theta\alpha + (1-\theta)\beta \\ \mu_{-}(\theta) = \left(\frac{\theta}{\alpha} + \frac{1-\theta}{\beta}\right)^{-1} \end{cases}$$

et de l'ensemble K_{θ} défini par

Si $0 < \theta < 1$, K_{θ} est l'ensemble des $\lambda_1, \dots, \lambda_N$:

$$(12) \begin{cases} \text{a)} \ \sum_i \frac{1}{\lambda_i - \alpha} \leqslant \frac{1}{\mu_{-}(\theta)-\alpha} + \frac{N-1}{\mu_{+}(\theta)-\alpha} = \frac{\theta}{(1-\theta)\alpha} + \frac{N}{(1-\theta)(\beta-\alpha)} \\ \text{b)} \ \sum_i \frac{1}{\beta-\lambda_i} \leqslant \frac{1}{\beta-\mu_{-}(\theta)} + \frac{N-1}{\beta-\mu_{+}(\theta)} = \frac{\theta-1}{\theta\beta} + \frac{N}{\theta(\beta-\alpha)} \end{cases}$$

$$(13) \quad \mu_{-}(\Theta) \le \lambda_i \le \mu_{+}(\Theta) \qquad \forall\, i\,.$$

En utilisant la convexité des fonctions $\{\varphi(\lambda) = \frac{1}{\lambda-\alpha}$ pour $\lambda > \alpha\}$ et $\{\psi(\lambda) = \frac{1}{\beta-\lambda}$ pour $\lambda < \beta\}$ on voit que K_{θ} est convexe.

Cet ensemble K_{Θ} va nous permettre de caractériser A^o.

Théorème 1. (Murat-Tartar) Si A^{ε} vérifie (10) et $A^{\varepsilon} \xrightarrow{H} A^o$ alors les valeurs propres $(\lambda^o_1, \dots \lambda^o_N)$ de la matrice A^o vérifient (14)

$$(14) \qquad (\lambda^o_1(x) \dots \lambda^o_N(x)) \in K_{\theta(x)} \quad \text{p.p.}$$

Réciproquement si une matrice $A^o \in (L^{\infty}(\Omega))^{N^2}$ symétrique a ses valeurs propres vérifiant (14) pour une fonction θ telle que $0 \leqslant \theta(x) \leqslant 1$ p.p. alors il existe une suite A^{ε} vérifiant (10) qui H-converge vers A^o. ∎

Ce théorème caractérise les matériaux qu'on peut fabriquer à partir de deux matériaux isotropes de caractéristiques respectives α et β . Il dit deux choses ; 1) une condition nécessaire : quelle que soit la répartition géométrique des deux constituants au niveau microscopique, les valeurs propres de la matrice effective A^o sont soumises à des contraintes dépendant des proportions utilisées. 2) une condition suffisante : on peut, en exhibant une géométrie particulière construire chacun des matériaux décrits par (14).

La condition nécessaire résultera d'une utilisation de lemmes de compacité par compensation ; la condition suffisante de calculs explicites dans le cas d'ellipsoïdes emboités ou plus simplement dans le cas feuilleté.

<u>Remarque</u>

Si $\int_\Omega \theta(x)\, dx = \gamma$ on peut imposer à la suite A^ε vérifiant (10) de satisfaire à

$$\int_\Omega C^\varepsilon(x)\, dx = (\alpha-\beta)\ \gamma + \beta \text{ mes } \Omega \,. \ \blacksquare$$

III . <u>Lemmes de compacité par compensation</u>

On se donne deux suites de matrices $\mathbb{P}^\varepsilon$, Q^ε vérifiant

$$(15)\quad \begin{cases} \mathbb{P}^\varepsilon \longrightarrow \mathbb{P}^o & \text{dans } L^2(\Omega)^{N^2} \text{ faible} \\ \forall i,j,k & \dfrac{\partial \mathbb{P}^\varepsilon_{ij}}{\partial X_k} - \dfrac{\partial \mathbb{P}^\varepsilon_{ik}}{\partial X_j} \in \text{compact } H^{-1}_{loc}(\Omega) \end{cases}$$

$$(16)\quad \begin{cases} Q^\varepsilon \longrightarrow Q^o & \text{dans } (L^2(\Omega))^{N^2} \text{ faible,} \\ \forall i & \sum_j \dfrac{\partial Q^\varepsilon_{ij}}{\partial X_j} \in \text{compact } H^{-1}_{loc}(\Omega) \,. \end{cases}$$

On a alors les lemmes suivants, où on note par $M^\star$ la transposée de la matrice M et par tr M sa trace :

<u>Lemme 1</u>

Sous les hypothèses (15) (16) on a

$$(17)\quad \forall i,k \quad \sum_j \mathbb{P}^\varepsilon_{ij}\, Q^\varepsilon_{kj} \longrightarrow \sum_j \mathbb{P}^o_{ij}\, Q^o_{kj} \quad \text{au sens des distributions.}$$

Lemme 2

Sous l'hypothèse (15) on a pour toute fonction φ vérifiant

$\varphi \in \mathcal{D}(\Omega)$, $\varphi \geq 0$

(18) $\liminf \int_\Omega \varphi[\mathrm{tr}\, \mathbb{P}^{\varepsilon\star}\, \mathbb{P}^{\varepsilon} - (\mathrm{tr}\, \mathbb{P}^{\varepsilon})^2]dx \geq \int_\Omega \varphi[\mathrm{tr}\, \mathbb{P}^{o\star} \cdot \mathbb{P}^{o} - (\mathrm{tr}\, \mathbb{P}^{o})^2]\, dx$.

Lemme 3

Sous l'hypothèse (16) on a pour toute fonction φ vérifiant

$\varphi \in \mathcal{D}(\Omega)$, $\varphi \geq 0$

(19) $\liminf \int_\Omega \varphi[(N-1)\mathrm{tr}\, Q^{\varepsilon\star} Q^{\varepsilon} - (\mathrm{tr}\, Q^{\varepsilon})^2]dx \geq \int_\Omega \varphi[(N-1)\mathrm{tr}\, Q^{o\star} Q^{o} - (\mathrm{tr} Q^{o2}]dx.$

Démonstration

On commence par calculer l'ensemble caractéristique :

(20) $\Lambda = \{(\mathbb{P},Q), \exists\, \zeta \neq 0 : \mathbb{P}_{ij}\, \zeta_k - \mathbb{P}_{ik}\, \zeta_j = 0\ \forall i,j,k\ ;\ \sum_j Q_{ij}\, \zeta_j = 0\ \forall i\,\}$

c'est à dire

(21) $\Lambda = \{\mathbb{P},Q : \mathbb{P} = a \otimes \zeta\ ,\ Q\,\zeta = 0$ pour un $\zeta \neq 0\,\}$.

La théorie générale [16] dit alors que pour toute fonction φ vérifiant

$\varphi \in \mathcal{D}(\Omega)$, $\varphi \geq 0$ on a

$\liminf \int_\Omega \varphi F(\mathbb{P}^{\varepsilon},Q^{\varepsilon})dx \geq \int_\Omega \varphi\, F(\mathbb{P}^{o},Q^{o})dx$ pour toute fonction quadratique F telle que $F(\mathbb{P},Q) \geq 0 \quad \forall(\mathbb{P},Q) \in \Lambda$.

Le lemme 1 découle alors du fait que $Q\,\mathbb{P}^\star = 0$ sur Λ .

Les lemmes 2 et 3 découlent alors du fait que rang $\mathbb{P} \leq 1$ et rang $Q \leq N-1$ sur Λ et du lemme d'algèbre linéaire suivant :

Lemme

Si M est une matrice carrée de rang $\leq r$ alors on a $r\ \mathrm{tr}\, M^\star M \geq (\mathrm{tr}\, M)^2$.

qui se voit aisément si les r premiers vecteurs de base engendrent l'image de M .

IV . Conditions nécessaires sur A^o :

La méthode [15] consiste à remarquer que d'après (7), (8) (en multipliant $\mathbb{P}^{\varepsilon}$ par une matrice fixe X) on sait construire deux suites $\mathbb{P}^{\varepsilon}$, Q^{ε} vérifiant (15) (16) avec $\mathbb{P}^{o} = X$ et $Q^{o} = A^{o} X$;
comme les oscillations de $\mathbb{P}^{\varepsilon}$ et Q^{ε} sont non seulement contraintes par (15) (16) mais aussi par la relation $Q^{\varepsilon} = A^{\varepsilon}\, \mathbb{P}^{\varepsilon}$ on voit que la connais- des oscillations sur A^{ε} va entrainer des contraintes sur A^{o} .

Plus précisément soit F une fonction, à croissance au plus quadratique vérifiant

(22) $\begin{cases} \liminf \int_\Omega \varphi\, F(\mathbb{P},Q)\,dx \geqslant \int_\Omega \varphi\, F(\mathbb{P}^o,Q^o)\,dx \quad \forall \varphi \in \mathcal{D}(\Omega)\ \varphi \geqslant 0 \\ \text{pour toute suite } (\mathbb{P}^\varepsilon,Q^\varepsilon) \text{ vérifiant (15), (16).} \end{cases}$

On définit alors une fonction d'une matrice A par

(23) $g(A) = \operatorname{Sup}_{\mathbb{P}} F(\mathbb{P}, A\,\mathbb{P})$.

Théorème 2

Si $A^\varepsilon \xrightarrow{H} A^o$ et

(24) $g(A^\varepsilon) \longrightarrow h$ dans $L^\infty(\Omega)$ faible ★

alors on a

(25) $g(A^o(x)) \leqslant h(x)$ p.p.

Remarquons que $g(A)$ peut valoir $+\infty$ pour certains A et que (24) contient donc des contraintes sur $A^\varepsilon(x)$.

Démonstration

On a $g(A^\varepsilon(x)) \geqslant F(\mathbb{P}^\varepsilon, A^\varepsilon\,\mathbb{P}^\varepsilon) = F(\mathbb{P}^\varepsilon, Q^\varepsilon)$

Donc $\int_\Omega \varphi\, h\, dx \geqslant \liminf \int \varphi\, F(\mathbb{P}^\varepsilon,Q^\varepsilon)\,dx \geqslant \int_\Omega \varphi\, F(\mathbb{P}^o, A^o\,\mathbb{P}^o)\,dx$

Donc $h(x) \geqslant F(\mathbb{P}^o, a^o\,\mathbb{P}^o)$ pp et comme $\mathbb{P}^o = X$ est arbitraire on en déduit $h(x) \geqslant g(A^o(x))$ p.p. ■

Si $g(\alpha I) < +\infty$ et $g(\beta I) < +\infty$ alors $h = \theta\, g(\alpha I) + (1-\theta)\, g(\beta I)$.

On peut en principe, appliquer la méthode à des situations plus générales mais le choix des bonnes fonctions F n'est pas clair : ici tout découlera des propositions suivantes.

Proposition 1

Si $F(\mathbb{P},Q) = \alpha\,[\operatorname{tr} \mathbb{P}^\star \mathbb{P} - (\operatorname{tr} \mathbb{P})^2] - \operatorname{tr} \mathbb{P}^\star Q + 2 \operatorname{tr} \mathbb{P}$ alors si A est une une matrice symétrique vérifiant $A \geqslant \alpha I$ (de valeurs propres λ_j) on a

$g(A) = \dfrac{\nu}{1+\nu\alpha}$ avec $\nu = \sum_j \dfrac{1}{\lambda_j - \alpha}$.

Démonstration

On se place dans une base diagonale de A et on doit maximiser en $\mathbb{P}$:

$$\alpha \sum_{i,j} P_{ij}^2 - \alpha(\sum_i P_{ii})^2 - \sum_{i,j} \lambda_i P_{ij}^2 + 2 \sum_i P_{ii} .$$

On voit qu'il faut prendre $P_{ij} = 0$ pour $i \neq j$ et le bon choix est ensuite $P_{ii} = \dfrac{\delta}{\lambda_i - \alpha}$ avec δ bien choisi : cela fait apparaître la quantité ν et on trouve facilement la valeur du maximum ■

Proposition 2

Si $\tilde{F}(\mathbb{P},Q) = (N-1)\ \mathrm{tr}\ Q^\star Q - (\mathrm{tr}\ Q)^2 - \beta(N-1)\ \mathrm{tr}\ \mathbb{P}^\star Q + 2\ \mathrm{tr}\ Q$ alors

si A est une matrice symétrique vérifiant $A \leqslant \beta I$(de valeurs propres λ_j) on a $\tilde{g}(A) = \dfrac{\tilde{\nu}}{\tilde{\nu}+N-1}$ avec $\tilde{\nu} = \sum_j \dfrac{\lambda_j}{\beta - \lambda_j}$.

Démonstration

Comme précédemment mais le bon choix pour $\mathbb{P}$ a la forme $P_{ii} = \dfrac{\delta}{\beta - \lambda_i}$ et $P_{ij} = 0$ pour $i \neq j$. Cela fait alors apparaître $\tilde{\nu}$

Grace aux lemmes 1 et 2 on peut appliquer le théorème 2 à la fonction g calculée à la proposition 1. [Cela donne $A^o \geqslant \alpha I$ car sinon on aurait $g(A^o) = +\infty$]. Comme $g(\alpha I) = \dfrac{1}{\alpha}$ et $g(\beta I) = \dfrac{N}{\beta+(N-1)}$ on obtient

$$\frac{\nu^o}{1+\nu^o \alpha} \leqslant \frac{\theta}{\alpha} + \frac{N(1-\theta)}{\beta+(N-1)\alpha} \quad \text{ce qui donne} \quad \nu^o \leqslant \frac{\theta}{(1-\theta)\alpha} + \frac{N}{(1-\theta)(\beta-\alpha)}$$

soit (12)a.

Grace aux lemmes 1 et 3 on peut appliquer le théorème 2 à la fonction $\tilde{g}$ calculée à la proposition 2 [cela donne $A^o \leqslant \beta I$ car sinon on aurait $\tilde{g}(A^o) = +\infty$]. Comme

$\tilde{g}(\alpha I) = \dfrac{N\alpha}{(N-1)\beta+\alpha}$ et $\tilde{g}(\beta) = 1$ on obtient

$$\frac{\tilde{\nu}^o}{\tilde{\nu}^o+N-1} \leqslant \theta \frac{N\alpha}{(N-1)\beta+\alpha} + (1-\theta) \quad \text{ce qui donne}$$

$$\tilde{\nu}^o \leqslant -N+1+\frac{\alpha+(N-1)\beta}{\theta(\beta-\alpha)} \quad \text{soit} \quad (12)\text{b} .$$

Remarque: Les conditions (13) sont classiques. On peut les obtenir à partir des choix $(F(\mathbb{P},Q) = -\mathrm{tr}(\mathbb{P}^* Q\ e \otimes e) + 2\ \mathrm{tr}(\mathbb{P}\ e \otimes e)$ qui donne $G(A) = (A^{-1} e,e)$ et $\tilde{F}(\mathbb{P},Q) = -\mathrm{tr}(\mathbb{P}^* Q\ e \otimes e) + 2\ \mathrm{tr}(Q\ e \otimes e)$ qui donne $\tilde{g}(A) = (A\,e,e)$.

V . CONDITIONS SUFFISANTES SUR A^o

- Soit $\theta \in L^\infty(\Omega)$ vérifiant $0 \leq \theta(x) \leq 1$ p.p. et $\int_\Omega \theta(x)dx = \gamma$.
  Soit $A^o \in (L^\infty(\Omega))^{N^2}$ symétrique dont les valeurs propres vérifient $(\lambda_1(x), \lambda_N(x)) \in K_{\theta(x)}$ p.p.
  On veut construire une suite $A^\varepsilon = c^\varepsilon(x)\, I$ qui H-converge vers A^o avec c^ε ne prenant que les valeurs α ou β , c^ε convergeant dans $L^\infty(\Omega)$ faible faible ★ vers $\theta\alpha + (1-\theta)\beta$ avec la contrainte $\int_\Omega c^\varepsilon(x)dx = (\alpha-\beta)\gamma + \beta \text{ mes } \Omega$.

- La première étape est de se ramener au cas où, sur un ouvert ω , θ et A^o sont constants.
  Compte tenu du caractère local de la H-convergence on saura traiter le cas où Ω est réunion d'un nombre fini d'ouverts disjoints (à un ensemble négligeable près) où θ et A^o sont constants.
  Compte tenu du fait que sur l'ensemble décrit par (2) la H-convergence provient d'une topologie métrisable moins fine que la topologie $(L^1(\Omega))^{N^2}$ fort on saura traiter le cas des limites fortes de (θ, A^o) constants par morceaux.

- Pour approcher (θ, A^o) par des fonctions constantes par morceaux il suffit de remarquer que l'ensemble M_θ des matrices symétriques dont les valeurs propres appartiennent à K_θ est convexe et que la distance de M_{θ_1} et M_{θ_2} est de l'ordre de $|\theta_1 - \theta_2|$.

  On approche d'abord θ par une suite θ_n constante par morceaux, de même intégrale que θ , et on prend pour $A^n(x)$ la projection de $A^o(x)$ sur $K_{\theta_n}(x)$.

  Sur l'un des ouverts Ω_j où θ_n est constant on approche facilement A^n par une suite constante par morceaux en utilisant le fait que M_θ est convexe.

<u>Définition</u> :
On dira qu'on peut fabriquer un matériau (de matrice) A à l'aide des matériaux (de matrice) A_j, $(j = 1,\ldots p)$ en proportions θ_j $(0 \leq \theta_j \leq 1, \sum_j \theta_j = 1)$ s'il existe une suite $\sum_j \chi_j^\varepsilon A_j$ qui H-converge vers A où les χ_j^ε sont les fonctions caractéristiques d'ouverts disjoints (pour le même ε)

vérifiant $\int_\Omega \chi^\varepsilon_j \, dx = \theta_j$ mes Ω et $\chi^\varepsilon_j \longrightarrow \theta_j$ dans $L^\infty(\Omega)$ faible ★ ■

Remarque

Il semble que la définition dépende de Ω, mais il n'en est rien : il suffit de voir qu'on peut déduire le cas général de celui où Ω est le cube unité. On décompose Ω en homothétiques du cube unité ; en plus du caractère local de la H-convergence on doit aussi utiliser l'effet d'une homothétie : si $\lambda \in \mathbb{R}_+$ et $y \in \mathbb{R}^N$ alors $a^\varepsilon(.) \xrightarrow{H} (a^o(.)$ implique

$a^\varepsilon(\lambda.+y) \xrightarrow{H} a^o(\lambda.+y)$ ■

- Utilisant une nouvelle fois la métrisabilité on voit que si A_j peut être fabriqué à partir de αI et βI avec les proportions η_j et $1-\eta_j$ alors le A ci-dessus peut être fabriqué aussi à partir de αI et βI avec les proportions θ et $1-\theta$ où $\theta = \sum_j \eta_j \theta_j$.

A ces considérations générales il faut maintenant ajouter des constructions plus explicites et nous utiliserons pour cela la formule des matériaux feuilletés :

Proposition 3

Soit $0 < \theta < 1$ et $e \in \mathbb{R}^N \setminus \{o\}$. A partir de A et B avec les proportions θ et $1-\theta$ on peut fabriquer C donné par

$$(26) \quad (C-A)^{-1} = \frac{(B-A)^{-1}}{1-\theta} + \frac{\theta}{1-\theta} \, \frac{e \otimes e}{(Ae,e)}$$

ou

$$(26)' \quad (C-B)^{-1} = \frac{(A-B)^{-1}}{\theta} + \frac{1-\theta}{\theta} \, \frac{e \otimes e}{(Be,e)} .$$

Remarque :

Si $B-A$ a un noyau $X \neq \{o\}$ alors on doit prendre (26) (26)' sur $X^\perp$ et compléter par $Cx = Ax$ sur X ■

Démonstration

Soit Ω le cube unité et χ_ε une suite de fonctions caractéristiques (en une variable) telle que $\int_\Omega \chi_\varepsilon(x.e) \, dx = \theta$ et $\chi_\varepsilon(x.e) \rightharpoonup \theta$ dans $L^\infty(\Omega)$ faible ★.

Posons

$$(27)\quad \begin{cases} E^{\varepsilon}(x) = \chi_{\varepsilon}(x.e)\, a + (1-\chi_{\varepsilon}(x.e))\, b \\ D^{\varepsilon}(x) = \chi_{\varepsilon}(x.e)\, Aa + (1-\chi_{\varepsilon}(x.e))\, Bb \end{cases}$$

où, $a,b \in \mathbb{R}^N$ sont choisis pour que (3) et (4) aient lieu :

(3) équivaut à (28) $\begin{cases} E^o = \theta\, a + (1-\theta) b \\ b = a + t\, e \quad \text{avec } t \in \mathbb{R} \end{cases}$

(4) équivaut à (29) $\begin{cases} D^o = \theta\, A\, a + (1-\theta) B\, b \\ (B\, b - A\, a.e) = 0 \end{cases}$

(5) a lieu avec $A^{\varepsilon} = \chi_{\varepsilon}(x.e)\, A + (1-\chi_{\varepsilon}(x.e)\, B$.

Montrons que (28) (29) donnent $D^o = C\, E^o$ avec C donné par (26) ou, en échangeant les rôles de A et B, (26)'.

Si $E_o \in \mathrm{Ker}(B-A)$ on prend $b = a = E_o$ et on a $D^o = A\, E^o = B\, E^o$.

Si $E_o \in (\mathrm{Ker}(B-A))^{\perp}$ on pose $\mu = (C-A)\, E_o$ avec C défini par (26),

$b = \dfrac{1}{1-\theta}(B-A)^{-1}\, \mu$ et $t = \dfrac{-(\mu.e)}{(1-\theta)(Ae.e)}$.

Alors on a bien $E^o = b - \theta\, t\, e$ à cause de (26) et $(Bb-Aa.e) = ((B-A)b.e) - t(Ae.e) = 0$ par construction. Il reste à calculer $D^o - AE^o = (1-\theta)(B-A)b = \mu = (C-A)\, E^o$ pour en déduire $D^o = C\, E^o$. ■

Proposition 4

Soit $0 < \theta < 1$; $\xi_j \in R_+ (j = 1,\ldots p)$ vérifiant $\sum_j \xi_j = 1-\theta$ et $e_1 ..,e_p \in \mathbb{R}^N \setminus \{0\}$. A partir de A et B avec les proportions θ et $1-\theta$ on peut fabriquer C donné par

$$(30)\qquad (C-B)^{-1} = \frac{(A-B)^{-1}}{\theta} + \frac{1}{\theta} \sum_j \xi_j \frac{e_j \otimes e_j}{(Be_j, e_j)}.$$

Démonstration

On part de $C_o = A$ et par récurrence on construit C_j à partir de C_{j-1} et B avec les proportions θ_j et $1-\theta_j$ en "feuilletant" perpendiculairement à e_j. La formule (26)' se prête bien à la récurrence et donne $\xi_j = \theta_1 .. \theta_{j-1}(1-\theta_j)$ ce qui permet de choisir les θ_j par récurrence. ■

Lemme

Si B est symétrique définie positive alors quand $\xi_j \geqslant 0$ $\sum_j \xi_j = 1-\theta$ et $e_j \in \mathbb{R}^N \setminus\{0\}$ la matrice $\sum_j \xi_j \frac{e_j \otimes e_j}{(Be_j, e_j)}$ parcourt toutes les matrices de la forme

$$(31) \quad \begin{cases} B^{-1/2} M B^{-1/2} \\ M \text{ symétrique semi définie positive et } \operatorname{tr} M = 1-\theta. \end{cases}$$

Démonstration

Posant $f_j = B^{-1/2} e_j$ on a $M = \sum_j \xi_j \frac{f_j \otimes f_j}{(f_j, f_j)}$

ce qui donne $M = M^\star \geqslant 0$ et $\operatorname{tr} M = \sum_j \xi_j$.

Réciproquement on choisit pour f_j une base diagonale de M et les ξ_j sont les valeurs propres de M.■

- Prenant alors $A = \alpha I$ et $B = \beta I$ on voit qu'on peut fabriquer avec les proportions θ et $(1-\theta)$ toutes les matrices C de valeurs propres $\lambda_1 \ldots \lambda_N$ vérifiant

$$(\lambda_j - \beta)^{-1} \geqslant \frac{1}{\theta}(\alpha-\beta)^{-1} \quad \text{et} \quad \sum_j (\lambda_j - \beta)^{-1} = \frac{N}{\theta}(\alpha-\beta)^{-1} + \frac{1-\theta}{\theta\beta}$$

c'est à dire $\lambda_j \leqslant \mu_+(\theta)$ avec égalité dans (12)b (ce qui implique $\lambda_j \geqslant \mu_-(\theta)$.

- En échangeant les rôles de α et β (et changeant θ en $1-\theta$) on obtient toutes les matrices C de valeurs propres λ_j vérifiant $\mu_-(\theta) \leqslant \lambda_j \leqslant \mu_+(\theta)$ avec égalités dans (12)a.
- Si maintenant on a inégalité stricte dans (12)a et (12)b ; en faisant varier λ_1 et en gardant $\lambda_2 \ldots \lambda_N$ constants on trouve un segment passant par le point cherché et dont les extrémités correspondent à des matrices C et C' qui vérifient l'égalité dans (12)a et (12)b et qu'on sait fabriquer. En mélangeant C et C' avec la proportion adéquate on obtient tous les points du segment, et on a donc ainsi montré qu'on obtient tous les points de K_θ.

VI . LA METHODE DES ELLIPSOIDES CONFOCAUX

Si la démonstration précédente (qui est une extension au cas A,B quelconques d'un calcul de Braidy-Pouilloux, analogue à celui de Lurie-Cherkaev, où A et B ont une base diagonale commune et les e_j sont les vecteurs propres) a l'avantage d'une certaine simplicité, la démonstration initiale (qui étend une construction de sphères emboitées utilisée par Hashin et Shtrikman) basée sur des calculs explicites pour des ellipsoïdes confocaux à d'autres avantages et vaut la peine d'être présentée aussi.

Soient $m_1 \dots m_N \in \mathbb{R}$; Soit B_ρ l'ellipsoïde dont le bord S_ρ a pour équation

$$(32) \qquad \sum_j \frac{x_j^2}{\rho+m_j} = 1 .$$

Pour $\rho + \operatorname{Inf}_j(m_j) > 0$ cela définit une fonction implicite

$$(33) \qquad \rho = \varphi_m(x)$$

définie en dehors d'une ellipsoïde dégénérée correspondant à $\rho + \operatorname{Inf}_j(m_j) = 0$.

Proposition 5

La fonction $\rho(x)$ définie ci-dessus vérifie

$$(34) \qquad \begin{cases} \dfrac{\partial\rho}{\partial x_k} = \dfrac{x_k}{\rho+m_k}\, \sigma_m(x) \\[2ex] \Delta\rho = \left(\displaystyle\sum_k \frac{1}{\rho+m_k}\right)\sigma_m(x) \end{cases}$$

$$\text{où } (35) \qquad \frac{1}{\sigma_m} = \frac{1}{2} \sum_j \frac{x_j^2}{(\rho+m_j)^2} .$$

Démonstration

Posons $\quad \psi_m(x) = \sum_j \dfrac{x_j^2}{(\rho+m_j)^2}$.

Dérivons (32) en x_k : $- \dfrac{\partial\rho}{\partial X_k} \sum_j \dfrac{X_j^2}{(\rho+m_j)^2} + \dfrac{2X_k}{\rho+m_k} = 0$

ce qui donne la première équation de (34) : $\dfrac{\partial\rho}{\partial X_k} = \dfrac{2}{\psi_m} \dfrac{X_k}{\rho+m_k}$.

Posons $\chi_m = \sum_j \frac{X_j^2}{(\rho+m_j)^3}$ et dérivons ψ_m en X_k :

$$\frac{\partial\psi m}{\partial X_k} = \frac{2X_k}{(\rho+m_k)^2} - 2\ \chi_m \frac{\partial\rho}{\partial X_k} = \frac{2X_k}{(\rho+m_k)^2} - 4\ \frac{\chi_m}{\psi_m}\frac{X_k}{\rho+m_k}\ .$$

Ensuite $\frac{\partial^2\rho}{\partial X_k{}^2} = \frac{2}{\psi_m}\frac{1}{\rho+m_k} - \frac{2}{\psi_m}\frac{X_k}{(\rho+m_k)^2}\frac{\partial\rho}{\partial X_k} - \frac{2}{\psi_m{}^2}\frac{X_k}{\rho+m_k}\frac{\partial\psi}{\partial X_k}$

c'est à dire $\frac{\partial^2\rho}{\partial X_k^2} = \frac{2}{\psi_m}\frac{1}{\rho+m_k} - \frac{8}{\psi_m{}^2}\frac{X_k^2}{(\rho+m_k)^2} + \frac{8\chi_m}{\psi_m^3}\frac{X_k^2}{(\rho+m_k)^2}$

d'où $\Delta\rho = \frac{2}{\psi_m}\sum_k \frac{1}{\rho+m_k}$ qui est la deuxième équation de (34).■

Remarque

On tire aussi de (34) (35) la relation utile

(36) $|\mathrm{grad}\,\rho|^2 = 2\,\sigma_m$.

Corollaire

Si (37) $g_m(\rho) = \prod_k \sqrt{\rho+m_k}$

alors $u = X_j\, f(\rho)$ est solution de $\Delta u = 0$ si

(38) $f'(\rho) = \frac{C}{(\rho+m_j)\ g_m(\rho)}$.

Démonstration

$$\Delta u = 2\, f'(\rho)\,\frac{\partial\rho}{\partial X_j} + X_j\left[f'(\rho)\,\Delta\rho + f''(\rho)\,|\mathrm{grad}\rho|^2\right],$$

et en mettant $X_j\ \sigma_m(X)$ en facteur on voit que $\Delta u = 0$ équivaut à

$$2\, f'(\rho)\,\frac{1}{\rho+m_j} + f'(\rho)\sum_k \frac{1}{\rho+m_k} + 2\, f''(\rho) = 0$$ qui donne

(38) où C est une constante arbitraire .

Remarque

Il y a aussi des solutions de $\Delta u = 0$ de la forme $f(\rho)$ avec $f' = \frac{C}{g_m(\rho)}$ mais elles ne nous serviront pas .

On considère maintenant deux ellipsoïdes de la même famille $B_{\rho-} \subset B_{\rho+}$ avec

$\rho_- < \rho_+$; on cherche à résoudre $-\text{div}(a \text{ grad } u) = 0$ où $a(x)$ vaut α dans B_{ρ_-} et β dans $B_{\rho_+} \setminus B_{\rho_-}$. D'après le corollaire il y a une solution de la forme $u = X_j f(\rho)$ si $f'(\rho) = 0$ pour $\rho < \rho_-$ et $f'(\rho) = \dfrac{C}{(\rho + m_j) g_m(\rho)}$ pour $\rho_- < \rho < \rho_+$ et si, à l'interface $\rho = \rho_-$, f est continue ainsi que $a \dfrac{\partial u}{\partial n}$ (où n est la normale de l'ellipsoïde, c'est à dire $n_k = \sqrt{\dfrac{\sigma_m}{2}} \dfrac{X_k}{\rho + m_k}$).

On trouve alors $\dfrac{\partial u}{\partial n} = X_j \sqrt{\dfrac{\sigma_m}{2}} \left(2f'(\rho) + \dfrac{f(\rho)}{\rho + m_j}\right)$

En imposant $f = 1$ pour $\rho < \rho_-$ on trouve $C = \dfrac{\alpha - \beta}{2\beta} g_m(\rho_-)$ et, pour

$$\rho_- < \rho < \rho_+ : f(\rho) = 1 + \frac{\alpha - \beta}{2\beta} g_m(\rho_-) \int_{\rho_-}^{\rho_+} \frac{1}{(\sigma + m_j) g_m(\sigma)} d\sigma .$$

On a donc, pour $\rho = \rho_+$, $a \dfrac{\partial u}{\partial n} = \dfrac{X_j}{\rho_+ + m_j} \sqrt{\dfrac{\sigma_m}{2}} \left[\dfrac{(\alpha - \beta) g_m(\rho_-)}{g_m(\rho_+)} + \beta f(\rho_+)\right]$

Si on avait rempli tout l'ellipsoïde $B_{\rho+}$ par un matériau de matrice $\tilde{a}$ ayant les valeurs propres $\lambda_1 \ldots \lambda_N$ dans les directions des axes alors $v = \gamma x_j$ vérifierait $-\text{div}(\tilde{a} \text{ grad } v) = 0$ et $\tilde{a} \text{ grad } v . n =$

$= \lambda_j \gamma \sqrt{\dfrac{\sigma_m}{2}} \dfrac{X_j}{\rho_+ + m_j}$ en $\rho = \rho_+$.

Si on choisit $\gamma = f(\rho_+)$ et $\lambda_j f(\rho_+) = \dfrac{\alpha - \beta}{g_m(\rho_+)} g_m(\rho_-) + \beta f(\rho_+)$

on voit que les deux problèmes correspondent aux mêmes données de Dirichlet et Neumann en $\rho = \rho_+$. La formule pour λ_j s'écrit

$$(39) \quad \frac{1}{\lambda_j - \beta} = \frac{1}{\alpha - \beta} \frac{g_m(\rho_+)}{g_m(\rho_-)} + \frac{g_m(\rho_+)}{2\beta} \int_{\rho_-}^{\rho_+} \frac{1}{(\sigma + m_j) g_m(\sigma)} d\sigma .$$

Si les λ_j vérifient (39) et si le matériau $\tilde{a}$ est soumis à un champ uniforme (parallèle à l'un des axes) on peut remplacer un ellipsoïde B_{ρ_+} par le mélange de deux ellipsoïdes emboités utilisant les matériaux αI et βI sans changer le champ extérieur. Par homogénéité on peut changer ρ_- ; ρ_+, m_j en $t\rho_-, t\rho_+, tm_j$ ce qui revient à changer la taille des ellipsoïdes, le rapport des volumes restant fixe avec

$\theta = \dfrac{g_m(\rho_-)}{g_m(\rho_+)}$. On peut donc par ce procédé fabriquer le matériau de valeurs propres λ_j.

Proposition 6

Si λ vérifie (39) alors pour $\theta = \dfrac{g_m(\rho_-)}{g_m(\rho_+)}$ on a

$$(40) \qquad \sum_j \frac{1}{\lambda_j - \beta} = \frac{N}{\alpha-\beta}\,\frac{1}{\theta} + \frac{1-\theta}{\theta\beta}$$

Réciproquement si $\mu_-(\theta) < \lambda_j < \mu_+(\theta)$ vérifient (40) il y a un choix de ρ_-, ρ_+, m_k qui donne les λ_j par la formule (39).

Démonstration

$\sum_j \int_{\rho_-}^{\rho_+} \dfrac{d\sigma}{(\sigma+m_j)\,g_m(\sigma)}$ se calcule facilement si on remarque que

$\dfrac{g'_m(\sigma)}{g_m(\sigma)} = \dfrac{1}{2} \sum_j \dfrac{1}{\sigma+m_j}$ et donc la somme vaut $2\left(\dfrac{1}{g_m(\rho_-)} - \dfrac{1}{g_m(\rho_+)}\right)$.

Pour montrer la réciproque on normalise par un choix de $\rho_- = 0$ et $\rho_+ = 1$ et on pose pour $m_j > 0$

$$(41) \qquad X_j(m) = \prod_k \sqrt{1+m_k} \int_0^1 \frac{d\sigma}{(\sigma+m_j)\prod_k \sqrt{1+m_k}}$$

et il suffit de montrer que quand m parcourt $(\mathbb{R}_+)^N$ alors $X(m)$ parcourt aussi $(\mathbb{R}_+)^N$.

La formule (40) équivaut ici à (42) $\sum_j X_j(m) = -2 + 2 \prod_k \sqrt{\dfrac{1+m_k}{m_k}}$.

Les X_j étant décroissants en m_k on fait un changement en m_k^{-1}.

Alors $Y(m) = X(m^{-1})$ a la propriété d'être continue de $(\overline{\mathbb{R}_+})^N$ dans lui-même, d'envoyer l'infini à l'infini et d'envoyer chaque face du bord dans elle-même (c'est à dire $m_j = 0$ pour $j \in J$ implique $Y_j = 0$ pour $j \in J$).

L'invariance par homotopie du degré topologique appliquée à l'homotopie $\varepsilon\,\text{Identité} + (1-\varepsilon)Y$ permet de montrer que pour $p \in (\mathbb{R}_+)^N$ on a $\deg(\Omega, Y, p) = 1$ où $\Omega = (\mathbb{R}_+)^N$ et donc qu'il existe $m \in \mathbb{R}_+^N$ avec $Y(m) = p$.

Le cas isotrope correspond au cas où les deux ellipsoïdes sont des sphères

la formule donne alors $\frac{N}{\lambda-\beta} = \frac{N}{\alpha-\beta}\,\frac{1}{\theta} + \frac{1-\theta}{\theta\,\beta}$ qu'on peut réécrire

$$\frac{\lambda-\beta}{\lambda+(N-1)\beta} = \theta\,\frac{\alpha-\beta}{\alpha+(N-1)\beta} \tag{42}$$

formule qui apparaît sous des noms divers (Maxwell-Garnett, Lorentz-Lorenz, Clausius-Mossoti) en physique pour approcher le coefficient effectif d'un matériau obtenu à partir du matériau β dans lequel on a inclus une proportion (petite) θ du matériau α .

BIBLIOGRAPHIE

[1] BERGMAN D.J. Bulk physical properties of composite media à paraître. Eyrolles : collection de la Direction des Etudes et recherches d'E.D.F.

[2] BRAIDY P. - POUILLOUX D. : Mémoire d'option, Ecole Polytechnique 1982 (non publié).

[3] DE GIORGI E. - SPAGNOLO S. : Sulla convergenza degli integrali dell' energia per operatori ellitici del secondo ordine. Boll. U.M.I., 8, (1973) p. 391-411.

[4] HASHIN Z. - SHTRIKMAN S. : A variational approach to the theory of effective magnetic permeability of multiphase materials. J. Applied phys. 33, (1962) p. 3125 - 3131.

[5] KOHN R.V. - STRANG G. : Structural design optimization, homogenization and relaxation of variational problems. Macroscopic properties of disordered media p. 131-147. Lecture Notes in Physics 154. Springer 1982.

[6] LURIE K.A. - CHERKAEV A.V. : Exact estimates of conductivity of composites. A.F. loffe physical technical institute, report 783. Leningrad 1982.

[7] LURIE K.A. - CHERKAEV A.V. : Optimal structural design and relaxed controls. Optimal contral Appl. , Meth. vol. 4 (1983) p. 387 - 392.

[8] MARINO A. - SPAGNOLO S. : Un tipo di approssimazione dell'operatore $\Sigma D_i A_{ij} D_j$. Con operatori $\Sigma D_j b D_j$. Ann. Scu. Norm. Pisa. 23, (1969). p. 657-673.

[9] MURAT F.:H-Convergence. Séminaire d'analyse fonctionnelle et numérique de l'Université d'Alger. 1977-78, ronéotypé 34 p.

[10] MURAT F. - TARTAR L. : Calcul des variations et homogénéisation, à paraître. Eyrolles : Collection de la Direction des Etudes et Recherches d'E.D.F.

[11] SPAGNOLO S. : Sul limite delle soluzioni di problemi di Cauchy relativi all' equazione del calore. Ann. Scu. Norm. Pisa. 21 (1967) p. 657-699.

[12] SPAGNOLO S. : Sulla convergenza di soluzioni di equazioni paraboliche ed ellitiche. Ann. Scu. Norm. Pisa. 22, (1968) p.577-597.

[13] TARTAR L. : Problemes de contrôle de coefficients dans des équations aux dérivées partielles. p. 420-426. Lecture notes in Economics and Mathematical Systems 107. Springer, 1974.

[14] TARTAR L. : Quelques remarques sur l'homogénéisation, p. 469-481. Japan, France Seminar Tokyo and Kyoto, 1976. H. Fujita ed. Japan Society for the promotion of Science 1978.

[15] TARTAR L. : Estimation de coefficients homogénéisés p. 364-373. Computing methods in applied sciences and engineering. Lecture notes in Mathematics 704. Springer, 1978.

[16] TARTAR L. : Compensated compactness and applications to partial differential equations, p. 136-212. Non linear analysis and mechanics : Heriot-Watt symposium vol. IV. Research notes in mathematics 39. Pitman 1979.

Luc TARTAR
Centre d'Etudes de Limeil-Valenton
94190 - VILLENEUVE SAINT-GEORGES
France

E DE GIORGI

Remerciements

Messieurs les Ambassadeurs, Madame le Recteur, Monsieur le Président, chers Collègues:

Je m'excuse si ma très médiocre connaissance de la langue française ne me permet de dire que quelques mots pour exprimer ma reconnaissance et celle de mes collègues qui viennent de recevoir la laurea honoris causa de cette Université de Paris, qui a été pendant des siècles et reste toujours un des plus remarquables centres de rayonnement culturel pour l'Europe et le monde entier.

Je pense que la cérémonie d'aujourd'hui est un signe de la volonté constante de l'Université de Paris de développer la coopération et l'amitié entre les scientifiques de différents pays et différentes disciplines. Cette amitié et cette coopération ont été toujours deux facteurs importants de progrès pour la science, une contribution remarquable à la compréhension entre tous les peuples, à la coopération internationale, à la paix et au progrès de l'humanité.

La conscience de ces vérités a été toujours vive dans la science et la culture françaises et je peux témoigner moi-même avoir appris de mes collègues de Paris non seulement bien des renseignements et des idées importantes pour mon travail en tant que mathématicien, mais aussi beaucoup d'exemples importants de la responsabilité des scientifiques envers l'humanité, de la liaison qui doit exister entre le progrès scientifique et la défense des droits de l'homme.

Le présent remerciement a été prononcé par E. DE GIORGI à la Sorbonne le 7.11.84, au nom des sept savants nommés ce jour là Docteurs Honoris Causa de l'Université de Paris.

Un historien, ou un philosophe, pourrait illustrer la contribution donnée par la culture française au développement de l'idée des droits de l'homme. Moi, je me bornerai à rappeler la valeur exemplaire qu'a eu pour moi et pour tous les mathématiciens du monde l'oeuvre du Comité des Mathématiciens, né en défense de Leonid PLIOUTCH et dès alors constamment engagé en défense impartiale de tous les mathématiciens du monde qui sont persécutés à cause de leurs opinions, qui ne peuvent pas exprimer librement leurs idées, qui ne peuvent pas rencontrer librement leurs collègues de tous les pays, qui ne peuvent pas participer au travail de la communauté scientifique internationale.

En particulier, aussi au nom de mes collègues de Pise, actuellement engagés dans la défense des mathématiciens José-Luis MASSERA et Joseph BEGUN, je dois rappeler que le Comité des Mathématiciens a toujours été un point de référence fondamental pour notre action. Au côté du Comité des Mathématiciens je désire mentionner les comités analogues des physiciens, des chimistes et des biologistes dont l'engagement est un témoignage important d'une vérité fondamentale : le progrès de la science, la défense des droits de l'homme, la paix, la liberté, la justice sont des biens inséparables et doivent être constamment recherchés par toutes les personnes qui aiment la vérité et le progrès de l'humanité.

Mais, outre la valeur de la solidarité entre les scientifiques de diverses disciplines, des divers pays, cette cérémonie doit nous rappeler aussi la solidarité idéale entre les scientifiques de tous les temps. Toute la science que nous possédons est le résultat du travail de générations de savants qui à diverses époques et dans divers pays ont servi les idéaux fondamentaux de la science, la recherche désintéressée de la vérité, le désir de communiquer leurs idées à tous ceux qui désirent les connaître, l'espoir que les découvertes scientifiques servent au bien de l'humanité et ne soient pas exploitées pour sa destruction. Je pense que chaque professeur et chercheur scientifique, afin de comprendre la valeur de son travail, doit le considérer comme partie de cette oeuvre millénaire de l'humanité, qui est un signe remarquable de la dignité de l'homme, de sa soif de connaissance que je crois être le signe d'un désir secret de voir quelques rayons de la gloire de Dieu.

Conscients de travailler à cette oeuvre commune, nous pouvons trouver la valeur et le dignité de la science, la coopération et l'amitié entre tous les scientifiques du monde, pour le bien de tous les hommes et de tous les peuples, manifester notre solidarité envers tous les hommes et tous les peuples du monde, surtout les plus faibles, les plus pauvres et opprimés, contribuer à préparer pour l'humanité un futur de progrès et de paix.

E DE GIORGI

References

Born 8 February 1928 in Lecce (Italy).

- Thesis on measure theory, with Professor Mauro PICONE, 1950 (University of Rome).
- Professor at the University of Messina, 1958-1959, and from 1959 at the Scuola Normale Superiore di Pisa.

Ennio DE GIORGI has contributed to the following mathematical theories.

I - Evolution problems

DE GIORGI proved a uniqueness theorem for parabolic differential operators (the Cauchy problem) whose coefficients belong to certain Gevrey classes. He gives a counter-example showing the nonuniqueness of solutions if the coefficients are less regular. In a joint work with A. MARINO and M. TOSQUES, DE GIORGI gives a general theory of evolution problems of variational type. This theory has recently been applied to the study of certains problems of nonlinear analysis.

II - Minimal surface

Starting from an idea of R. CACCIOPOLI, DE GIORGI introduced new ideas in the geometric theory of measurement and in applications to variational calculus. He proved a theorem of almost-everywhere regularity for minimal surfaces. These results are presented in a book written in collaboration with F. COLOMBINI and L.C. PICCININI. Later, in a joint work with BOMBIERI, MIRANDA and GIUSTI, he proved the regularity of minimal surfaces and gave a counter-example showing that Berstein's analycity result cannot be extended to dimensions greater than 7.

III - Regularity of solution of second-order elliptic partial differential equations

The Hölder-continuity of solutions of equations of this type has been proved by E. DE GIORGI (and also independently by Nash). E. DE GIORGI found an example with discontinuous solutions. These results, combined with earlier ones, give the solution to D. Hilbert's problem conerning the analyticity of extremals for multiple integrals.

IV - Analytic solutions of partial differential equations with constant coefficients

The existence of such solutions was first proved in two dimensions by E. DE GIORGI in collaboration with L. CATTABRIGA. This result has later been generalized by several authors.

V - Gamma-Convergence and connected problems

The first result in this field was proved by E. DE GIORGI and S. SPAGNOLO (1973) for second order elliptic differential equation. Later E. DE GIORGI elaborated a general theory of Gamma-Convergence with important applications to the direct methods of the calculus of variation, and also to homogenisation problems in mechanics, probability theory, geometry... The theory of Gamma-Convergence has been,and is still, the source of numerous works and the range of its application seems to be very wide.

VI - Hyperbolic equations with discontinuous coefficients with respect to the time variable

In a joint work with F. COLOMBI and S. SPAGNOLO, DE GIORGI proved an existence theorem for solutions of the Cauchy problem with analytical initial data. A counter-example shows how this theorem fails if the initial data are not analytic.

Published papers

I - Evolution problems

1) Un esempio di non unicità della soluzione del problema di Cauchy relativo ad un'equazione differenziale lineare a derivate parziali di tipo parabolico. Rend. di matem. 14 (1955), 328-387.

2) Un teorema di unicità per il problema di Cauchy, relativo ad equazioni differenziali lineari a derivate parziali di tipo parabolico. Ann. di Mat. pura ed appl. 40 (1955), 371-377.

3) Problemi di evoluzione in spazi metrici e curve di massima pendenza (with A. MARINO and M. TOSQUES). Atti Accad. Naz. Lincei Rend. Cl. Sci. Fis. Mat. Natur., (8) 68 (1980) 180-187.

4) Funzioni (p,q)-convesse. (with A. MARINO and M. TOSQUES). Atti Naz. Lincei Rend. Cl. Sci. Fis. Mat. Natur. to appear.

5) Evolution equations for a class of non-linear operators. (with M. DEGIOVANNI, A. MARINO and M. TOSQUES). Atti,Acc. Naz. Lincei, Rend. Cl. Sci. Fis. Mat. Natur. to appear.

II - Boundaries and minimal surfaces

1) Un nuovo teorema di esistenza relativo ad alcuni problemi variazionali. Communication at the III Conference of the Austrian Mathematical Society (9-15 September 1952). Report N. 371 of I.N.A.C.

2) Definizione ed espressione analitica del perimetro di un insieme. Atti Acc. Naz. Lincei, Rend. Cl. Sci. Fis. Mat. Natur. 14 (1953).

3) Su una teoria generale della misura (r-1)-dimensionale in uno spazio ad r dimensioni. Ann. di Matem. pura ed appl. 36 (1954), 191-213.

4) Nuovi teoremi relativi alle misure (r-1)-dimensionali in uno spazio ad r dimensioni. Ricerche di Matem. 4 (1955), 95-113.

5) Sulla proprietà isoperimetrica dell'ipersfera, nella classe degli insiemi aventi frontiera orientata di misura finita. Memorie Accad. Naz. Lincei 5 (1958), 33-44.

6) Una estensione del teorema di Bernstein. Ann. Scu. Norm. Pisa 19 (1965), 79-85.

7) Sulle singolarità eliminabili delle superficie minimali (with G. STAMPACCHIA). Atti Accad. Naz. Lincei, Re,d. Cl. Sci. Fis. Mat. Natur. (8) 38 1965, 352-357.

8) Una maggiorazione a priori relativa alle ipersuperfici minimali non parametriche (with E. BOMBIERI and M. MIRANDA). Archive for Rat. Mech. and Analysis 4 (1968), 255-267.

9) Minimal cones and the Bernstein problem (with E. BOMBIERI and E. GIUSTI). Inventiones Math. 7 (1969), 243-268.

10) Frontière orientate di misura minima e questioni connesse (with F. COLOMBINI and L.C. PICCININI). Pubblicazione della Classe di Scienze della Scuola Normale Superiore di Pisa (1972), 1-179.

III - Variational problems of elliptic type

1) Sull'analiticità delle estremali degli integrali multipli. Atti, Acc. Naz. Lincei, Rend. Cl. Sci. Fis. Mat. Natur. 20 (1956) 438-441.

2) Sulla differenziabilità e l'analiticità delle estremali degli integrali multipli regolari. Memorie Accad. Sci. Torino, Ser. III, Parte I, (1957), 25-43.

3) Un esempio di estremali discontinue per un problema variazionale di tipo ellittico. Boll. U.M.I., 1 (1968), 135-137.

4) Teoremi di semicontinuità nel calcolo delle variazioni. Seminario, Istituto Nazionale di Alta Mathematica, Roma 1968-69.

5) On the lower semicontinuity of certain integral functionals. (with G. BUTTAZZO and G. DAL MASO), to appear.

IV - Existence of solutions of differential equations with constant coefficients

1) Una formula di rappresentazione per funzioni analitiche in R^n (with L. CATTABRIGA). Boll. Un. Mat. It. 4 (1971), 1010-1014.

2) Una dimostrazione diretta dell'esistenza di soluzioni analitiche nel piano reale di equazioni a derivate parziali a coefficienti costanti (with L. CATTABRIGA). Boll. Un. Mat? It. 4 (1971), 1015-1027.

3) Sull'esistenza di soluzioni analitiche di equazioni a derivate parziali a coefficienti costanti in qualunque numero variabili (with L. CATTABRIGA) Boll. Un. Mat. It. 6 (1972), 301-311.